Cities in the Urban Age

Cities in the Urban Age

A Dissent

ROBERT A. BEAUREGARD

The University of Chicago Press   Chicago and London

The University of Chicago Press, Chicago 60637
The University of Chicago Press, Ltd., London

Published 2018
Printed in the United States of America

27 26 25 24 23 22 21 20 19 18   1 2 3 4 5

ISBN-13: 978-0-226-53524-1 (cloth)
ISBN-13: 978-0-226-53538-8 (paper)
ISBN-13: 978-0-226-53541-8 (e-book)
DOI: 10.7208/chicago/[9780226535418].001.0001

Library of Congress Cataloging-in-Publication Data

Names: Beauregard, Robert A., author.
Title: Cities in the urban age : a dissent / Robert A. Beauregard.
Description: Chicago ; London : The University of Chicago Press, 2018. | Includes bibliographical references and index.
Identifiers: LCCN 2017024308 | ISBN 9780226535241 (cloth : alk. paper) | ISBN 9780226535388 (pbk. : alk. paper) | ISBN 9780226535418 (e-book)
Subjects: LCSH: Cities and towns. | Cities and towns—Social aspects. | Urban economics. | Cities and towns—Political aspects. | Cities and towns—Social aspects—United States. | Cities and towns—Political aspects—United States.
Classification: LCC HT151 .B38 2018 | DDC 307.76—dc23
LC record available at https://lccn.loc.gov/2017024308

♾ This paper meets the requirements of ANSI/NISO Z39.48–1992 (Permanence of Paper).

Contents

Preface

With few exceptions, those who currently write and speak about the city do so in celebration. They portray the city as the driving force of national economies—a hinge that connects countries to global flows of capital, people, and ideas. The city nurtures innovation, brings forth unlimited opportunities to pursue fame and fortune, and, with its high density and large-scale infrastructures, offers the best hope for environmental sustainability. To live in this city is to be surrounded by a vast and rich array of culinary, cultural, and consumer pleasures. There, people from different backgrounds, national origins, and lifestyles create a vibrant and a tolerant public realm. And, while some people might choose to live elsewhere, no country can prosper without such cities.

At least this is what we are told.

One would have to be mean-spirited to deny to those who make this argument their fascination with the city. I, too, am endlessly intrigued and, at times, even in awe. What bothers me, though, is the subtext of progress. The underlying premise is that, over the centuries, the city has become better and better at serving the needs of the planet. Today's boosters would have us believe that humans have slowly but relentlessly devised a form of human settlement that offers security and prosperity while holding forth the possibility of surmounting the mundane demands of everyday life. For such urbanists, the claim that more than one-half of the world's population now resides in cities is less an empirical observation than a thinly veiled, triumphal boast. National fanaticism, wars, religious intolerance, and institutionalized corruption might resist all of our efforts to eliminate them, but the city ostensibly exists as the one human achievement that has become increasingly adept at meeting our needs.

As two well-known policy experts recently wrote: "The metropolis is humanity's greatest collective act of invention and imagination." They are not alone in their opinion: the economist Edward Glaeser has declared that cities "enable the collaboration that makes humanity shine most brightly." More equivocally, but no less enthralled, Nan Rothschild and Diane diZerega, both archaeologists, claim that cities "are arguably the most significant of human inventions." Such laudatory, even hyperbolic, statements are not only prevalent in the contemporary literature on the city but generally uncontested as well.[1]

The city, and specifically the US city—the topic of this book—was not always so inviting. In fact, throughout much of its history it has been under the sway of contradictory impulses with prosperity existing alongside impoverishment and tolerance entwined with prejudice and discrimination. In short, the city is not wholly advantageous. As recently as the early twentieth century, its residents were overwhelmed by epidemics, were made ill by contaminated water, lacked basic health care, and had limited access to education. Most people of the time condemned the city as a pestilence and abomination that signaled the decay of civilization, not its realization. The cities' residents, noted one observer in 1911, "lived in smoke, amid ugly and incongruous buildings with unattractive highways, often poor and almost always inadequate, and without suitable parks, park space, trees, and other aesthetic essentials."[2] The city seemed more of a burden than a blessing.

In the mid-twentieth century, these same cities—all former manufacturing centers—entered a period of unrelenting and chronic loss of population and jobs. Soon thereafter, deprived of tax revenues, their local governments began to reduce the public services that had kept the city clean and well maintained and its residents protected from harm. Conditions deteriorated even further when African-Americans rioted in protest of police brutality and the dismal conditions under which they were forced to live. For some, these years of "urban crisis" marked the death of cities and presaged their eventual extinction. The future seemed to lie in less dense forms of human settlement—the mass-produced suburbs in particular. A generalized aversion to cities—an anti-urbanism—became part of the national culture in many industrialized countries.[3] Yet, a mere fifty years later, the city has been rescued from this dire fate and has ostensibly become—once again—a symbol of human civilization and a cultural achievement of immense significance.

This widely shared and flattering view rests on the presupposition that the city exists and endures because it allows us to share lives of prosperity, opportunity, and fulfillment. As the world's population has grown, innovations have proliferated, economies have expanded, and people have been drawn together on the land. Across the centuries, we are told, the city has emerged as the settlement form that best enables people to provide for their needs and nurture their aspirations. The city seems to solve many of the problems that people confront in finding a place to live, raising families, creating communities, forming societies, and inventing nations. This is what progress looks like.

Praise is always comforting, and the point of view that gives rise to this flattery is so prevalent and so embedded in how we think about the city that it is impossible to imagine a more appropriate way of bringing people and activities together. What would be the alternative? Any replacement would seem to require a radical change in how people engage with each other, the many ways they meet their needs, and how they wish to be governed. Such an alternative borders on science fiction. We can no more imagine a world without large cities than we can imagine a future without smartphones, automobiles, and surveillance cameras.

Contemporary urban commentary is rooted in this assumption of progress, of a successive overcoming of challenges and a corresponding improvement in how we live together. New construction techniques, medicines, means of communication, production processes, and commodities mark the advance of civilization. This assumption of progress, however, is wrong.

The city has a dark side, acknowledged by even its most rabid advocates, of concentrated poverty, slums, racial discrimination, environmental destruction, anti-immigrant sentiment, and (now) terrorism. Too many residents in too many cities are exploited, marginalized, or treated unjustly. Although we have made advances in disposing of human waste, providing housing for large numbers of people, protecting the natural environment, and maintaining public health, thereby making it possible for large and dense cities to function, many of the problems these advances address—contaminated water, the paucity of inexpensive and high-quality housing, for example—persist. Moreover, the benefits have not accrued to everyone. Neither have contemporary cities banished the injustices and inequalities that plague all human relations. Nor are they ever likely to do so.[4]

The city does not erase the conflicts that arise when people live in close proximity: it does not eliminate criminal behavior, insensitivity toward others, the urge to dominate, or the quest to amass personal wealth to the detriment of others. The city's many positive qualities contribute in important ways to prosperity, innovation, and social solidarity. And, yet, evil and vice, destitution and alienation, and corruption persist. Antagonisms abound with wealth and poverty, tolerance and intolerance, environmental stewardship and degradation joined together in city after city. In fact, by providing fertile ground in which such antagonistic forces can flourish, the city enables these contradictions to endure. The city is not a solution to life's dilemmas; it enables them.[5]

The city is a crucible. At one and the same time it serves multiple interests, virtues, vices, and passions and pits diverse publics in opposition to each other. The city's history is one of recurring antagonisms and unstable compromises. Contradictions are ever-present—never surmounted, only renegotiated time and again. The result is not progress but rather an abiding incompleteness. Cities do not "survive by adapting to challenges" as if these challenges appear, are confronted, and are summarily dispensed.[6] Challenges persevere. One only has to survey the city's history to appreciate how malleable and unsettled it can be in the face of demographic changes, technological innovations, revolutions and wars, industrial collapse, and natural disasters. The city accommodates and amplifies, repels and dampens. It heightens or diminishes the antagonisms among groups as it grows and declines and as people come into more frequent and more unavoidable contact. The famous urbanist Jane Jacobs recognized this quality. She wrote about and celebrated cities as inefficient and impractical—difficult to manage—and proclaimed that this is "exactly what makes cities uniquely valuable."[7]

We cannot understand the city without attending to these paradoxical qualities. The city is the ground on which society's contradictions are contested. While it does not create these contradictions, it mediates between them, sometimes nurturing often amplifying, and, at times, even dampening their opposing forces. More than "a place of extremes," as if those extremes could be bridged, the city is "fundamentally disordered."[8]

To give substantive weight to this argument, I have organized the book around four contradictions: that between wealth and poverty, environmental destruction and sustainability, oligarchy and democracy, and intolerance and tolerance.[9] Together, they capture the challenges and choices that confront those who live and work in the city.

First, although cities make great wealth possible, they also concentrate it in the hands of a few individuals, families, and corporations. This gives rise to poverty and in almost all cities we find stark disparities in income and wealth and, following from that, visible discrepancies in living conditions. These inequalities are more pronounced and widespread in cities than in small towns and rural villages. While they vary in extent from one city to the next, they are never fully absent.

Second, cities are the most environmentally destructive form of human settlement yet devised, but they also hold out the best hope for environmental sustainability. Their extensive ecological footprints are a consequence of their large size, complexity, and high density, all of which are essential for technologies that conserve energy and protect animal and plant habitats. Neither low-density suburban developments nor small towns have the capacity to house the world's expanding human population without irreparably damaging the environment.

Third, cities are ideal places for democracy to thrive. When people live close together and have common needs, they are compelled to share in the work of governance. Yet, cities enable political influence to be concentrated among a few people in ways that displace and weaken democracy. They offer numerous opportunities for those with private wealth or political connections to exploit public resources and affect governmental policy. Consequently, local governments are almost always more oligarchic than democratic. That said, the conditions created by these oligarchic tendencies frequently (although not inevitably) engender popular resistance aimed at reasserting democracy.

Fourth, while cities encourage people to be tolerant, they also nurture intolerance. To live easily in the city, one has to allow others to be different. For centuries, observers have celebrated the city's public spaces for their openness and people have flocked to cities to live in ways that some might consider idiosyncratic and even inappropriate. In the competition for neighborhoods, jobs, and housing, however, differences often breed antagonism, contempt, and violence. Under such conditions, people experience hate crimes, ethnic and religious groups face discrimination, and people of color are segregated in housing and labor markets. Although tolerance exists, it does so precariously.

Widely acknowledged, these four contradictions haunt urban life. Even observers who celebrate the city concede their existence. Poor families, abandoned industrial districts, crime, and environmental threats can be ignored but hardly wished away. At the same time, recognition has to be given to vibrant business districts, attractive parks, and comfortable neighborhoods as well as a robust politics, tolerance for diversity, and efforts to confront climate change. The problems of cities are not anomalies and their good qualities are not automatic. They almost always exist together.

What explains this? An obvious place to turn for an answer is the literature on Marxist urban political-economy. Dialectic thinking, with its emphasis on the tensions that stalk society, is fundamental to the worldview of Marxist scholars. At its root are the many ways in which capitalist societies empower capitalists to exploit labor, drive down wages, and create marginalized neighborhoods ravaged by disinvestment. Control over space enables the rich to maintain the value of their homes and keep others from enjoying the amenities—the good schools, playgrounds, and food retailers—they have gathered around them.

Central to this Marxist, urban perspective is the idea that societal contradictions have their roots in processes that transcend city boundaries. Cities are embedded in larger settings that are open to the ideas, people, goods, and money that flow among them. Investment capital and migrants incessantly cross political boundaries despite the efforts of national governments to slow or redirect these flows. New financial instruments are devised in New York City and adopted in the banks of Reykjavik, coal plants pollute the air in China and temperatures rise in Europe, and war erupts in Syria and the effects are felt in Berlin and Washington, DC.[10] For Marxist scholars, these contradictions originate in capitalism itself, a mode of production that imposes its antagonisms on the world.

Non-Marxist scholars also recognize contradictions and paradoxes but attribute them instead to the scarcity of resources or the venality of humans. Rare is the non-Marxist urban scholar who fails to mention "urban problems." In doing so, however, they almost always treat such problems as temporary malfunctions of an otherwise advantageous city, not as contradictions that are central to it. For the most rabid of celebrants, the city's problems are an afterthought to its ability to generate wealth and attract creative people.[11]

Contradictions, then, are neither transitional states in the city's full development nor correctable consequences of human fallibility. If they were, we would see clear evidence that people learn from their experiences and, so doing, eliminate the malfunctions, conflicts, acts of hubris, and greed that make the city inhospitable for too many of its inhabitants. The city being celebrated is also cast as solely a human achievement, as if humans acted alone without the need of technologies such as water pumps, electricity, and telephones and without having to work with nature as embodied in water tables, edible plants, and air inversions. The city's contradictions are neither temporary nor reducible to human behavior—and they are unavoidable.

To be clear, the city is a setting for these contradictions; it does not create them. To make the latter credible, I would have to convince you that the world is divided between places suffering all of the consequences of such dilemmas, and places devoid of them, as if one could flee St. Louis for the suburbs or for the hills of southwest Missouri and, by doing so, travel to where the only antagonisms are personal and fleeting. Safely ensconced beyond the city's limits, life would be free of irresoluble conflicts—almost edenic. Regrettably, the world is not so neatly arranged.

Blaming the city for these contradictions would also mean treating it as if it has intentions and preferences. The city is not like a person or a firm, a government agency, or even a river. The city does not do things. It does not act "as one." Rather, the city is an array of people and activities that are constantly grouping and regrouping, influencing one another in myriad ways, and seldom acting harmoniously or with a single, underlying intent. It facilitates a vast number of actions and innumerable consequences, only a portion of which are predictable or even immediately perceived. Consequently, we cannot write that the city creates wealth or that by becoming too big or too congested it produces its own decline. The city does no such thing; it is the setting and not the player.

In short, the city is inherently unsettled; it is always in flux as it relentlessly adapts to changing populations, technologies, lifestyles, and institutions. It never reaches equilibrium and so can never be judged complete or even celebrated for its accomplishments. Similarly, the city is also unbounded. We cannot understand Dallas by focusing only on what occurs inside of it. The city's influence extends well beyond its political limits and much of what Dallas is and how it functions originates from afar. Even attributing to it a fixed set of functions is questionable since the city serves a multitude of ever-changing needs and desires. Claiming that it has a single or even dominant function—to generate wealth, for example—is simply wrong. Yet, many who write about the city focus on only one of its many qualities.[12] They present it as a collection of neighborhoods, a control center for regional and global economic relations, or a place of shared governance and often highlight a single, noteworthy attribute: the technological stature of Palo Alto or the religious significance of Salt Lake City. The decision to do so is understandable. The city can never be comprehended or described in all of its complexity. But, there is danger in reducing its complexity and subduing its dynamism. To conceive of the city solely as a mechanism for generating wealth, an incubator for human creativity, or merely a physical form does it a disservice. We can never portray, or even know, the many aspects of the city, but neither should we then opt for simplification. What we should do instead is recognize its contradictory nature.

Emphasizing these four contradictions is more than my attempt to influence how to think about and perceive the city. It is also a statement about how we position ourselves morally in relation to one another. If we conceive of the city as a product of human progress, then harmony and prosperity become inevitable. In such a world, no dilemmas have to be resolved and no enduing conflicts stand in the way of a better future. That is not the city we live in, however. When we define the city in terms of its oppositions and paradoxes, the innumerable choices that affect and define our responsibilities to each other and to nature become unavoidable. The city's complexity and unsettledness bring forth a constant stream of occasions when we are compelled to assess the consequences of opposing tendencies that deepen inequalities and disrupt ecological communities while, at other times, reduce social disadvantages, protect the environment, and enhance democracy. Rather than being swept into the future by progress, such that injustices, suffering, and marginalization are successively eliminated, we are confronted by endless moral choices. As a result, we are compelled to become political beings who acknowledge the differences among us and, in recognition and of necessity, assume civic responsibility. That not everyone does so, or acts accordingly, is simply another contributor to the city's contradictions.

For illustration, I draw on evidence from cities in the United States. I know those cities best. This is not to imply that my argument makes sense only in this one country; it can be applied fruitfully to cities in other countries and cultures as well. Likewise, my examples will mainly reference current events and conditions; this is not a history. I want you to sense the contemporary city in its substantive details and to consider my perspective alongside an abundance of examples.

My hope is that the reader will finish with a better understanding of how cities function and a greater appreciation for what it means to live together with people different from one's self and one's group. In addition, I hope that this approach to thinking about Tucson, Mobile, Burlington, and Lansing dispels the prevailing ethos; celebration of the city marks a retreat not just from the unsettledness that makes the city so attractively vibrant but also from the miseries and conflicts that are all too present and for which we are responsible.[13]

Each generation decides how it will position itself in relation to the city. The current generation of commentators has elected to celebrate it, and most of the population of the United States and the world has chosen or been compelled to live within its domain. A prior generation in the United States, contemplating the death of the city, was not so forgiving. I have written this book for the next generation. I hope that many of them too will be attracted to cities and find living there satisfying and stimulating. Notwithstanding, they should never lose sight of the city's contradictory tendencies or the complex and ever-changing relations that are its nature. If the next generation thinks of the city as simply another marker of human progress, I will have failed. To hold this conviction is to cling to a worldview at odds with reality. The city is malleable, but not infinitely so. It is accommodating, though only to a degree. It is unsettled, but not to the point of chaos. It is susceptible to modification and even rejection. The arrangements that people make to live well together and prosper are always provisional. The city, the type of human settlement that currently looms large in human affairs, might well be only temporary. That it is incessantly perplexing and endlessly fascinating makes me want to believe otherwise.

ONE

The City

Until the 1850s, a family dissatisfied with village life in Massachusetts might have relocated to the farmlands of Ohio or further northwest to the woods of Michigan. It would not have felt the need to live in a city; life in its new home would have been perfectly fine. Cities did exist. One was quite large with nearly 400,000 residents, but aside from New York City, places like Austin, Louisville, and Baltimore neither were big enough nor had become so entangled with the rest of the country as to dominate the choices that people made in deciding where to live. Subsistence in villages and small towns, on the plains and in the forests, was no less satisfying because it was beyond the city's reach. Not yet an "urban" nation with people clustered together in dense settlements, the United States was still a country where living beyond the limits of the city in no way diminished one's quality of life.

Between the late nineteenth and the mid-twentieth centuries, city residence became more and more unavoidable. The mechanization of agriculture and the consolidation of farming, the shift of the economy to mass-produced goods and the rise of large factories, the shrinking of distances brought on by the telegraph and the railroad, and increasing interdependencies generated by bank financing had a profound effect on the country's settlement patterns. Certain cities came to dominate national life and the city extended its influences deeper and deeper into areas heretofore untouched. Unless one lived near or within a city, doing well was more and more difficult for large numbers of people. New forms of transportation (particularly trolleys and, later, the automobile) eventually enabled people to reside beyond urban borders where development was less dense, nature more accessible, and the city's noise, congestion, and social diversity at a distance. The residents of these peripheral places, however, had not escaped the city's influence.

Today, slightly more than 8 of every 10 households in the United States reside in or directly adjacent to a city. Access to jobs, education, cultural opportunities, wealth, and even political careers are increasingly found in core cities and their adjacent suburbs, what together are called metropolitan areas. The non-urban possibilities of the past century have almost wholly disappeared. Even people who carry on a solitary existence high in the Rocky Mountains or deep in Florida's Everglades cannot break from the city's reach. To be totally alone, one has to be self-sufficient and "off the grid," not just disconnected from the technologies of modern life but also severed from government and the many benefits that it offers.

In the early years of this century, then, it has become virtually impossible to be untouched by what happens in cities. One can still decide whether to work or live there, but the decision to establish a home in the suburb of Marietta rather than in Atlanta city proper still positions one's family amid that city's many influences. The suburbs where most adults commute to work are simply extensions of the core. Locating further beyond the urban fringe seldom means leaving behind all of the things of the city: health care, magazines and television, smartphones and the internet. A family can live far outside the municipal boundaries of Butte or El Paso; it is not, however, outside of it in a way that makes the city irrelevant to daily life.

Cities, however, are not just a functional framework for human existence, a machine for living. They also have an imaginative component. When popular commentators and politicians reflect on what it means to be an American, the city is inevitably drawn into the conversation. Debates about immigration, public versus for-profit (charter) schools, climate change, mass transit, and tax policy all hinge on basic considerations about how we arrange ourselves on the land in relation to others, how we coexist with nature, and what technologies we deploy to allow people to live closer together or further apart. Do we go to the movies, a diversion produced for cities, or hike in the foothills? Do we imagine a single-family home surrounded by a well-trimmed lawn and replanted shade trees or imagine coming home in the evening to a cozy apartment on the fourteenth floor of a building indistinct from its neighbors? Do we drive to work on country lanes or descend subway platforms? Do we mingle with strangers or mostly encounter friends and acquaintances as we move through the day? These and numerous other questions arise as we consider how we want to live and with whom. All of them encourage us to consider whether the city resonates with our sense of who we are and that of the groups to which we belong. Are we comfortable being ourselves in these surroundings? Are we able to live with others who provide us with support?

From its European beginnings, the country did not lack what might have been then considered the equivalents of cities. Plymouth Colony in what is now Massachusetts, Jamestown in Virginia, and New Amsterdam, the Dutch trading port that eventually became New York City, were (after awhile) relatively large, relatively dense settlements that provided for many, though hardly all, of the daily needs of their inhabitants. Early on, various items—cloth, furniture, tea—still had to be imported from England, the Netherlands, and France, but, as settlements grew, more and more goods were made locally. Compared to contemporary cities like Seattle and Nashville, colonial cities hardly deserve the label. By current standards, they were small and thin on the ground. The activities that occurred there were rudimentary and not at all diverse, businesses were few, and public services nonexistent. The work that people did was quite similar from one person to the next. This changed as more and more European settlers arrived and towns and cities became not just bigger, denser, and more complex but also more numerous as well. Chicago went from a trading post of 200 residents in 1833 to a large, diverse, and complex city of 3.6 million in 1950. Las Vegas, a mere 25 people in the desert in 1900, grew to over one-half million by the end of the century. By 2014, the United States had 34 cities with more than 500,000 inhabitants and 10 cities and 62 metropolitan areas of over one million residents. Today, its cities, including their suburban extensions, contain the great majority of the country's population.[1]

Cities have so taken hold of life in the United States that it is near-impossible to imagine a world without them. Across the long span of human history as the number of people on the planet has grown to an estimated 7.5 billion, more and more of them have chosen or been compelled to live in densely populated settlements. This process of urbanization—the expansion in the number and sizes of cities—seems almost inexorable.[2] Cities have become the dominant form of human settlement and a significant presence in countries around the world. What does it mean to live in the city? How does doing so expand and even, at times, narrow the gap between how we want to live—individually, in families, amid others like ourselves, as a nation—and how we actually do live? What is it that cities do?

These are not just rhetorical questions to be left unanswered, but central to the perspective of this book. My goal is to provide you with a way to think about the city such that you can imagine living there among its many paradoxical tendencies. So situated, I hope you will also reflect on the moral implications these contradictions pose.

Before addressing these issues, we need to reflect on what a city is. When households with young children leave Chicago for better schools in Evanston or Somali refugees relocate to Omaha, what are they leaving and where are they going? How are these new places different from those they were drawn to or have left behind? Most people have no problem with these questions. To a great extent, popular discussion allows the city to draw its meaning from the way in which it is used within a conversation. When a television news reporter announces a mass shooting in Charlotte, we know where they mean, more or less. For scholars and researchers, however, a bit more precision is required. The meaning of the city needs to be questioned. The same concern applies to my argument about cities being the crucible for society's contradictions. More needs to be made explicit. Three themes will provide the necessary background: First, whether we think of the city as a place or an object like a tree or offshore oil rig; second, how we distinguish a city from that which is not a city; and, third, once we have identified something called a city, the extent to which it can be considered stable, self-contained, and even knowable.

Thinking the City

For urban scholars, "the city" is an object of endless fascination, whether they treat it as a place where things happen—for example, where music genres take root as in Detroit in the early 1960s and the Motown Sound—or as a "thing-in-itself"; that is, as a coherent and singular object of interest as is often done with Washington, DC (a center of global politics) or Hollywood (the hub of the film industry). These distinctions matter. When considered as a place, our attention often turns to the people, organizations, and activities located within its boundaries and to how cities differ from each other. So, for example, we might consider how immigrants adapt differently in El Paso than in Chicago. Which city is more accommodating? Underlying this question is a very important assumption: where something happens affects how it happens. Geography and social relations are conjoined. Treating the city as a self-contained object, in contrast, turns our attention outward to the relations between and among cities. Prevalent in the contemporary literature is a discussion of how cities compete with each other for capital investment, international cultural events, and highly educated immigrants.[3] Implied is that the city is coherent and has integrity similar to that of an airplane or a book and acts much like a person or corporation. It is not obvious, however, that cities are self-contained and self-organizing objects. And, it might not even be a useful way to think of them. For a number of scholars, the focus should be placed instead on the processes of how cities come to be, not on the form they take. As one has commented: "We do our understanding of cities a great disservice when we focus on them as objects, rather than on the urbanization processes that produced them."[4]

Urban scholars are continually debating these possibilities. By embracing the multiplicity of the city, as we will do, its status as either a simple setting or as a thing-in-itself disappears. Identifying its effects on what happens there and the relationships it encompasses become problematic. In this alternative perspective, what we think of as urban extends "beyond the confines of the city itself." [5] Indicative of this approach is Jennifer Robinson's comment that the city is both a place and "a series of unbounded, relatively disconnected and dispersed, perhaps sprawling and differentiated activities, made in and through many different kinds of networks stretching far beyond the physical extent of the city."[6] Definitions such as Robinson's, however, seem to encourage a vagueness that is simply unhelpful, particularly in a literature saturated with discussions of the city that never quite define what is being discussed.

To talk casually about cities, then, is easy; to define them in a way that absorbs all of their complexity and mystery is not. A commentator might deploy a simple definition—the city as any dense human settlement—and then drape it with qualifications, treating the definition as merely an opening statement. Or, she might cast the city in such abstract terms—the city as the physical embodiment of the human spirit—that one can never usefully connect the definition to actual places and activities. My inclination is to do the former; that is, to build up an understanding of the city from a very rudimentary, relatively concrete statement to a more elaborate understanding.

I begin with a basic statement that characterizes the city as a geographical concentration of diverse people and activities that is considered by most observers to be relatively large and relatively dense in terms of people and functions. This is a very common definition and draws on the notion of agglomeration from economic geography and a classic article by the urban sociologist Louis Wirth that reflected on what it meant to be urban, a condition that he labeled *urbanism*.[7] For Wirth, urban places were large in size, comparatively dense, and heterogeneous in their inhabitants and activities. These are its defining elements. His use of the concept of urbanism was meant to focus attention on how people live within cities, their ways of life. In doing so, Wirth suggested that the city is not just a physical place but also a way of being in the world. To this extent, it is both experienced and perceived. When we discuss the city, then, we are pointing both to a material presence and to imaginative possibilities.

On the physical side, cities are places where people and activities have concentrated and done so increasingly over the centuries because of building technologies such as steel construction and infrastructures such as subway systems and water distribution networks. In the process, once-natural landscapes have become occupied but not wholly displaced. The physical city is a network of humans and their various forms of association (for example, families, ethnic groups); reservoirs, rats, and land forms by which nature is integrated into the city; and electricity systems, traffic regulations, and newspaper distribution mechanisms that represent the technologies without which people could not live together in such proximity. The city is a social, ecological, and technical system of relatively large size, density, and complexity. It contains humans and nonhumans, some of which (like pigeons) are living and others of which (like street lights) are not. To this extent, cities are not solely human achievements but rather achievements forged in collaboration with nature and technologies.

Life in cities also has an imaginative component. People live there and think about themselves as living there. Humans are self-aware and respond to the world through both physical cues and the meanings they attach to those cues. In short, their understandings of their experiences are part of those experiences. When we act, we think about what it means to act and how our actions are being perceived or might be perceived by others. To this extent, the city lives in the minds of those who engage with it and does so whether they confront it directly as users or vicariously through various media. Thus, while the city is not just a physical place, the physicality of the city influences how we think about it. This does not happen in any determinative way. Midtown Manhattan's office skyscrapers and crowded sidewalks enable us to imagine it either as vibrant and prosperous or as discomforting and threatening.[8]

This imaginative city does not just operate in the present. Through memory, it also invades the past. We remember what the city was like—Providence's downtown filled with shoppers headed for department stores and the Lower Ninth Ward of New Orleans before its devastation by Hurricane Katrina in 2005. These remembrances filter our perception of the current moment and influence discussions about whether to live there or not, visit or not. Such memories affect the stories told about cities along with collective decisions regarding the buildings and places to be preserved in order for their history to be celebrated. Decisions about what investments need to be made to recapture the past or move the city away from conditions stifling its development are similarly filtered. Deliberations about historic preservation, economic development, public capital investments, gentrification, and business retention—among many others—are filled with references both to the physical city (an eighteenth-century graveyard) and to imaginings of its future. Does the city's history require that the burial ground remain where it is or can it be relocated for a new concert hall?

People thus become attached to places within the city, not only because they have bought a home or a small industrial building for their computer repair business, but also because these places are part of how they imagine their heritage. Such imaginaries enrich their lives in the present. It matters to a Jewish group that its synagogue not be replaced by a high-rise condominium project and that the buildings where pickles were made, meats cured, and poultry sold not be torn down. A lost place is a memory tarnished and identification with place is an important aspect of being a city dweller. People imagine themselves as Bostonians, Portlanders, or citizens of New Orleans. This image contributes to who they are and attaches them to the city. Such identifications can be even more specific; we are Southies (as in South Boston), residents of Society Hill and not Chestnut Hill (as in Philadelphia), or Santa Monicans (as in the Los Angeles metropolitan area). And, these identifications can also become blurred as when a person whose home is in Somerville tells a business associate at an out-of-town convention that she lives in Boston or resides in Pasadena but tells a stranger from the East Coast that she lives in Los Angeles.

Attachments to place are most salient when deciding what to do when rents skyrocket, children leave for college, or the neighborhood becomes less safe. Then, people have to decide whether to stay or move elsewhere. Or, should they, instead, organize to change what is pushing them away? Would a new zoning regulation allow their now-empty bedrooms to be used to accommodate an in-law apartment? Would heightened police presence reduce crime? Whether they leave or voice their disapproval, the physical transformation of a place disrupts the meanings people have of it. With the future less predictable, their lives become unstable. Their attachments to place begin to weaken.

A number of commentators go even further to characterize cities as having different ethos or cultures.[9] San Francisco, for example has always been known for its tolerance of diverse and even transgressive lifestyles. Cambridge, just adjacent to Boston, is popularly viewed as a city of intellectuals: professors and students, universities, and research institutes. Seattle has developed an entrepreneurial reputation. At the same time, and exposing these characterizations to doubt, is that cities are not simple organisms (as discussed earlier) that can be so singularly described. Boston in the nineteenth century was considered quite staid, as was Philadelphia, at least in comparison to New York, but it has also had periods of intense political activity as well as racial conflict, particularly notorious in the 1960s when school desegregation was underway. Yet, there does seem to be something different in the spirit of Los Angeles as compared to Pittsburgh or New Orleans. New Orleans stands out because of its reputation as a place of good food, good times, and easy living. These cities are not just physically distinct but imaginatively distinct as well, even though we cannot precisely say what those imaginative differences are.

Keeping in mind that cities are both physical things and ideas (that is, existing in our minds), we are still left with the problem of having to distinguish the city from other forms of human settlement. Simply stating that cities are the largest, densest, and most complex of these communities—qualities that can be turned easily into numeric measures—suggests that a quantitative distinction is sufficient. Many urban scholars would disagree. For them, the city is qualitatively different from a suburb, a small town, a village, a hamlet, and the scattered farms and homes of rural dwellers. Consider this comment from Jane Jacobs' *The Death and Life of Great American Cities*, one of the best-known books in the urban studies library: "Towns, suburbs, and even little cities are totally different organisms from great cities."[10] What might this mean? How are cities different from non-cities and then, within the category of cities, different from each other? This is the second of this chapter's background issues.

Despite the skepticism that I voiced above, size, density, and diversity do play a role in making these distinctions. Places that have relatively few people living some distance from each other and lacking a full range of the goods, services, and activities necessary to live a modern life usually fail to qualify as cities. Such settlements might have only a few hundred residents living in detached houses each occupied by only one or two families. These houses are likely to be spread far apart and surrounded by fields or woodlands. Most importantly, while a resident can pick up the mail, buy gas, and purchase a loaf of bread a few miles distant, the place lacks a full-service supermarket, a lawyer's office, a medical clinic, and a shoe store. A person has to travel elsewhere to buy certain items or services and to consult a tax specialist or attend a professional sporting event. These settlements are not considered cities. Or, to reverse the thought, "big cities are not only larger and denser than suburbs and small towns, but are actually different, experientially richer and culturally and economically more diverse."[11]

Many cities were once non-cities. As they grew, they added residents and businesses, became denser as land became more expensive and developers and property owners made buildings higher and put them closer together, and became large enough in population for the residents to support a shoe store or an accountant. They came to be more and more like cities. When, though, does a growing settlement become a city and no longer a village or small town? The answer to this question can be reasonable and defensible, but is nevertheless arbitrary; no objective criteria allow us to distinguish a city from a non-city similar to how we might distinguish between being in the continental United States and having left it. The distinction is purely a matter of convention. Many US researchers work with government population data and use 50,000 residents as a dividing line between cities and non-cities.[12] But not only does this ignore density and diversity, it has no strong defense on either theoretical or practical grounds. Not much difference exists between a "city" of 52,000 people and a "town" of 48,000 people as regards either how land is used, peoples' daily lives, or how many different goods and services can be purchased there.

When scholars and popular commentators discuss cities and distinguish them from other types of places, they often refer to something beyond physical attributes. For most people, cities have a certain, even unique, feeling. This returns us to the imaginative dimension of cityness or, better, urbanity. New York, Seattle, Denver, and Washington, DC, feel like cities. They have skylines, sidewalks busy with pedestrians, streets packed with cars, delivery trucks, and taxis that assault us with a cacophony of horns and police sirens. People move with intent or they sit leisurely along the edges of plazas. One can find a place to enjoy a glass of microbrew beer, watch a foreign movie, attend a professional wrestling match, or browse through a comic-book store. Street vendors sell eyeglasses and religious groups proselytize. People are ethnically diverse and almost all unfamiliar—strangers. You are anonymous. The public spaces of Center City or of the Northern Liberties neighborhood in Philadelphia have the feel of a city. These are not villages or small towns.[13]

The "feel" of a city is palpable, even if almost impossible to define or measure. It seems to have something to do with size, density, and diversity but "what" eludes us. Even cities that meet such quantitative criteria are often denied the label. Not too many years ago, city lovers accused Los Angeles (then the third largest city in the country) of not being a city. It had no "center," its downtown streets were empty at night and its sidewalks barely used, and businesses, hotels, doctors' offices, and sports stadiums were scattered across the land rather than clustered in the core. Similar comments have been made about Phoenix, Houston, and Las Vegas. One commentator has written about Phoenix's downtown that "no one would confuse it with the core of a major urban center."[14] Where, these commentators ask, is the city? These places might be big, relatively dense, and multifunctional, but they are so fragmented spatially that they lack a sense of being a city. A visitor from Chicago or Baltimore would view them as imposters. Why do its residents and visitors have to travel to and fro across the city to find the activities that one expects there? Why is there no street life?

Such distinctions have a physical dimension that supports other notions of urbanity. For example, we label as cities both Boston with its dense concentration of people and activities, its skyline, its high property values, and its subway system, and Detroit with its acres of empty land, innumerable vacant and abandoned homes and factories, a faltering city government, and the predominance of poor, unemployed, and undereducated residents. Compared to Boston, Detroit seems to have lost its urbanity, despite having approximately 730,000 residents in 2014 and a downtown (even if one that is "dead" at night). At the same time, we are willing to accept Muncie, Indiana, or Worcester, Massachusetts, as being cities even though they are small places, are not very dense, and fall short of offering the array of goods and services one can find in Boston. The idea of the city is pulled and stretched across a variety of places.

Objective measures are of help, but urbanity, this particular sense of cityness, is something to be experienced rather than subjected to science. Sidewalk cafes, night life, entertainment venues, office towers, and active public spaces are certainly conditions that we can document and even measure, but their combination, along with certain fleeting qualities, is what makes a place feel like a city. We cannot really say, definitively, what a city is in these terms and thus are unable to draw a fine line between places where large numbers of people cluster together and cities. (Of course, we can always arbitrarily assert a distinction, as governments and researchers frequently do.) If there is a continuum, consensus is only available at the extremes. After that, it is all more or less open for discussion, as it should be.

A recent response to this problem is to reject any distinction between the urban and the non-urban. From this perspective, the city and the countryside are expendable as geographic categories. The argument is that the distinction inhibits clear thinking about the process of urbanization, not that Denver or Providence do not exist. There is no place, these theorists claim, that has not been penetrated by or come under the influence of an urbanization process that "transgresses, explodes, and reworks inherited geographies."[15] We live in a time of planetary urbanization. The flow of trade, the spread of music and fashion, advanced telecommunications from cell-phone towers to the internet, international migration, and air travel have connected the far corners of the earth. Every place has been touched by diplomacy, commercialization, email, property markets, and television. Nomadic tribes in the Jordanian desert have access to cell phones and their movements and livelihoods are affected by territorial disputes rooted in global conflicts. This argument resonates with those scholars who claim that capitalism, and not the city, is the vehicle for the colonization of the world.[16] Today, however, those arguing for "the urban" treat capitalism as only one aspect of this planetary phenomenon. Urbanization is the more encompassing and dominant force in both its material and imaginative manifestations. It is the force behind globalization.

Such an argument provides an answer to "what is a city?" by displacing the question. The issue is not cities as a material presence but urbanization as a process. To distinguish cities from non-cities, specifying what a city is, these theorists argue, only detracts attention from the central issue of how "the urban" has spread across the landscape such that no place escapes its influence. Yet, what is meant by *urban* is still unclear and here we encounter the same issues that we do with *city*. Neither objective measures nor intimations of urbanity provide a solution. Urbanization cannot be defined independently of the city and thus little is gained by shifting the terminology and ignoring the city's material presence as if stability or the physicality of the city does not matter. How do we know when a place has become a city, that is, has made the transition from relatively uninfluenced by urbanization to engulfed by it? To what extent does this argument assume the prior existence of unadulterated places where people lacked the qualities we now associate with "urban," an assumption that resonates with the notion of progress? The whole argument has bothersome overtones of the nineteenth-century distinction between civilization and its absence.

If we reject the notion of a fully urbanized world, and thus continue to believe in both cities and non-cities and their importance for how people live and corporations and governments act, we come up against the third, background issue. How do we distinguish between where a "real" and specific city—Tacoma—ends and something else begins? When we talk about Boise, Idaho, or Mobile, Alabama, to what "space" are we referring?

In the United States, cities are considered to be different from the communities that surround them—the suburbs.[17] The more precise term for the former is central cities for, historically, they were the centers of commerce, cultural life, and governance for their regions. Together, the central cities and their suburbs constitute metropolitan areas. In these places, the central city and its suburbs are interdependent; they are joined together by business relationships, the commuting of workers from one to the other, political bodies that manage water and transportation on a regional basis, identification with the professional sports teams or cultural offerings such as museums and symphonic orchestras, and the television, radio stations, and newspapers that serve the entire region. For many commentators, and particularly those concerned to capture the influence of cityness, the "real" city is the metropolis. One can no more separate the city from its suburbs than one can separate the wheels from a bicycle and still pedal to the neighborhood park. The interdependencies between them are so thick that it makes no sense to consider one in the absence of the other. The city is the metropolitan area.[18]

Once we accept this argument, then we are back to trying to identify where the influence of cities (or metropolitan areas) ends. Obvious, for example, is that commodity markets in Chicago affect the decisions that farmers in Iowa make about crops and livestock and thus the prosperity of these rural places.[19] And crop failures in Iowa are going to reverberate in commodity markets in Chicago. This muddles any notion of a functional urban (or, more precisely, economic) region, which is what a metropolitan area is meant to be. One solution is to claim that places must be contiguous to be part of the city, thereby preserving the notion of the metropolitan area. But this means ignoring the undeniable influence of cities beyond their metropolitan boundaries and diminishes our understanding of them. We do not want to adopt the position that the whole world is urban, but neither do we want to dismiss how cities have come to affect the country itself as well as places much farther away. Cities have consequences beyond their borders, whether those borders are drawn tightly around the central city or more loosely around the metropolitan region. Adding a good deal of confusion is that such functional borders hardly ever reflect political boundaries.[20]

Making these distinctions more difficult to navigate is that within the US federal system of government, cities have a political status that clearly distinguishes them and their territories from suburbs, small towns, and villages.[21] One can argue that Tampa has to be thought of as being its metropolitan area, but Tampa is nonetheless a municipality separate in tax capacity, residency laws, and governmental arrangements from the suburbs such as Brandon and Temple Terrace that surround it. The city has its own elected officials, school system, and streets department. A person who lives there, moreover, is considered an official resident of Tampa—the place where she votes and pay taxes—and not a citizen of the metropolitan region. Politically, Tampa is a distinct entity and is treated this way by both the Florida state government and the federal government. Regional agencies might coordinate waste management and highway construction, but for many issues such as affordable housing, public schools, building code enforcement, and policing, what matters to elected officials, property owners, and residents is the city itself and not the functional urban region. The city has a special political status that cannot be ignored, a fact that has stifled efforts to match political territories with functional territories. The metropolis might be the economic unit, but it is politically fragmented and, for researchers and scholars intent on knowing the city, this further complicates matters.

In some ways, when we imagine or speak of cities, the thing to which we refer is relatively clear. Spurred by tourism and media representations, most cities are associated with specific images and to mention those cities is to convey a relatively precise sense of place. New Orleans evokes the French Quarter with its numerous clubs and restaurants and the many revelers and bands that parade during Mardi Gras. Mention Philadelphia, and one imagines the Art Museum and the Benjamin Franklin Parkway along with the Liberty Bell and Independence Hall. Mention Miami and the images are of gleaming, balcony-clad apartment buildings overlooking deep-blue waters. In visualizing a city, we often focus on iconic objects and places: think the Space Needle in Seattle and the adobe buildings surrounding the plaza in Santa Fe. The distant view is often clearer than that on the ground. Referring to the 1970s, the historian Sam Bass Warner argued that the city was dominated by two images: the skyline and the ghetto.[22] Today, the skyline continues to represent cities in the popular imagination, but people are more likely to think in terms of vibrant streets lined with sidewalk restaurants and people strolling along waterfront promenades than of poor, dilapidated, and racialized neighborhoods.

Whether in the foreground or the background, the notion of the city is elusive. We can neither specify precisely what it is nor fully encapsulate it within our descriptions. Cities are restless, with their consequences extending beyond their political boundaries. To ask where New Orleans ends and other cities begin is to confront the city's multiplicity with a human need to bring order to the world. New Orleans exists in—and has consequences for—multiple places. After Hurricane Katrina in 2005, many of its residents fled to Houston and yet still maintained property and family relations in New Orleans. Representatives from the city travel to Washington, DC, to influence legislation on bayou restoration and shipping channels. Without connections to oil companies around the world, tourists from Chicago, Los Angeles, and London, and federal aid from the nation's capital, the city would wither. Additionally, New Orleans has an imaginative presence that reaches to jazz musicians in Boston and revelers celebrating Mardi Gras in Kansas City.

The city is also mysterious, existing just beyond our comprehension and not fully resolved. That does not mean that we cannot write and talk about it. In fact, we can measure its physical presence and study its changes, although doing so requires that we make educated but essentially arbitrary, though defensible, distinctions concerning the city and its boundaries. What is required for those who wish to present a broad understanding is that the city's meaning is clear when the term is used, even if that meaning changes as the city is portrayed from different perspectives. In such discussions, precision is overrated and ambiguity a friend.

Within this framework, I place the four contradictions that constitute the core of this book. The city is not fixed; it is not a thing. In addition, the city is unsettled and complex, and its workings elusive. It is diverse in the things that constitute it and impossible ever fully to comprehend. It is neither a single thing nor a solo actor in a play about redevelopment, ethnic conflict, or political corruption. Neither is a city like a person with a more or less coherent identity and intentions. What is important here is that thinking about the city in this way keeps us from turning the city into a fetish, an object of worship, and from attributing to it, and it alone, anthropomorphic qualities. The city is not an organism like a human body. And, it is not, as one observer has boasted, "the greatest social experiment in human history."[23] The critical issue for us is what it enables people to do and what it prevents them from doing. To emphasize again the theme of the book, this depends on its contradictions. No city, no one, can escape them.

Four Contradictions

For me and many others, the city is more than elusive and mysterious, complex and ever-restless; it harbors contradictions.[24] A contradiction exists when a corporate policy, an economic system, or an ideological agenda contains opposing tendencies. These tendencies are neither unintended nor self-canceling. They are not accidents; they do not balance or cancel each other out, creating a Panglossian "best of all possible worlds." Unavoidable and yet contingent, they cannot be eliminated and their consequences do not always manifest in the same way. National legislature, in order to balance contending interests, might fund programs that discourage people from driving automobiles (for example, placing a heavy tax on gasoline consumption) and, at the same time, encourage them to do so by building highways and skimping on the funding of mass transit. An industrialist might lower wages to increase the company's profits, but this also reduces the purchasing power of workers and thus the demand for what the company makes. Such contradictions are not inherently urban; nonetheless, they are pronounced in cities. Even if it does not create contradictions, the city nurtures them.

This quality of cities is often attributed to capitalism rather than urbanism.[25] The argument is that a specific economic arrangement is contradictory, not that the city is. Capitalism's contradictions involve a deeply rooted division between those who own the means of production (and thus the means of producing income and wealth) and those who do not. Karl Marx, capitalism's great theorist of the nineteenth century, noted the numerous ways in which opposing tendencies were harbored within the capitalist class, divided capital and labor, and stood between the workers and the state. He pointed out how the driving down of wages enabled capitalists to extract more profits but also weakened the health of the labor force needed to create these profits. Marx also pointed to how competition among capitalist firms hindered the kinds of collaborations that would provide public goods such as harbors or roadways without which raw materials could not be shipped to factories or finished goods to markets. In these instances, government has to step in to protect capitalism from itself.

Marx made a compelling argument for these opposing tendencies within capitalism and for locating society's contradictions in its political economy. For urban theorists such as Henri Lefebvre, however, the city is best seen as relatively independent of capitalism. The "urban"—conflicts over land use, the juxtaposition of diverse activities, the assembly of humans and ecology—is what gives society its dynamism and fuels its paradoxes.[26] In effect, Lefebvre subordinates capitalism to urbanism: "Urban reality modifies the relations of production. . . . It becomes a productive force."[27] With such comments, he suggests that cities, not capitalism, generate—one might say "cause"—the contradictions that haunt society. I would disagree. To resurrect and invert an old idea—geography is not destiny.

To be clear, I am not claiming that cities create contradictions that subsequently spread beyond their boundaries into the larger society. Rather, contradictions play out and originate across various spatial scales from the factory floor to the world and resist being geographically confined. That is, I am not portraying the city, as one urban theorist has warned, "as the locale of a myriad of social ills, as if these shortcomings were uniquely or inherently urban, and as if cities and their problems were not considered versions of the relations that characterize the society as a whole."[28] What I argue instead is that society engenders these conditions and then they are crystallized and mediated within cities. Because it is of the world, the city is complicit. But more than that, it articulates these contradictions in ways that give them significance for how we both understand and live in cities.

That said, four contradictions, broadly framed, are of concern: (1) the ability of cities simultaneously to generate both wealth and poverty, (2) the extent to which cities' environmental destructiveness is linked to the opportunities they hold for environmental sustainability, (3) the nurturing of democracy at the same time that cities engender various forms of oligarchy and oppression that concentrate political power, and (4) the many ways in which tolerance is encouraged even as various groups are marginalized and humiliated. These consequences articulate what most contemporary urban theorists and popular commentators take to be the defining characteristics of cities. They also allow me to range across the city's diverse manifestations and convey what cities do for the people and nonhuman things that inhabit them.

First, cities are robust mechanisms for creating both private and public wealth. More so than rural settlements, they enable people to earn large sums of money and amass property from expensive homes to corporate stocks and Impressionistic paintings. At the same time, they empower governments to build public parks, schools, commemorative monuments, recycling facilities, and bridges. The density and interdependencies of economic activity, the competitive pressures, the need for supportive infrastructure, and the public demand for health and well-being are only some of the factors contributing to this capacity for producing goods and services and consolidating monetary rewards. And, while these various mechanisms can falter and struggling cities present fewer and fewer opportunities for living well, the potential for generating wealth is almost always present, even if dormant.

On the other hand, cities do not harbor strong mechanisms for distributing wealth in an equitable manner among their residents. Wealth is frequently concentrated among a few families and groups rather than fairly shared. In cities, we find the country's richest corporations, its wealthiest citizens, expensive art galleries, and the most prestigious universities with the largest endowments. We are also likely to find high concentrations of people living in poverty, deteriorated neighborhoods, troubled schools, and areas where services such as health care and libraries are inadequate. For many of the country's wealthiest cities, the disparity between those who live exceedingly well and those who barely eke by is vast. The richest households earn many times what working-class households do and their wealth is often infinitely greater since the poorest of households have no wealth at all on which to rely. Poverty and inequality coexist with wealth and prosperity and this seems an unavoidable consequence, suggesting that, in contradictory fashion, these conditions are inseparable.

Second, cities have a contradictory relationship with nature. On the one hand, and particularly when compared with lower density settlements such as the typical suburb of detached, single-family homes of postwar vintage, cities are environmentally advantageous. Living in apartment buildings consumes less land. Traveling by bus or subway minimizes air pollution and uses less energy per person. Water and sewer infrastructure can be provided more efficiently, as can mail delivery. In these and many other ways, cities hold forth the potential to be sustainable or, more accurately, more sustainable than other forms of human settlement. Cities, though, are also environmentally destructive. Cities use up vast amounts of natural resources (especially land) and concentrate air, water, and soil pollution. To function well, moreover, cities have to extract, and be able to extract, resources from an area larger than that which they occupy. Their *ecological footprint*, as this is known, is much, much bigger than their territorial footprint. As cities grow in size and number, more and more of the planet is affected. This expansion has important consequences for ecologies, natural resources, and air and water quality around the world. By making cities more sustainable and more resilient, we encourage them to expand and, in doing so, inflict even more harm on the natural environment.

Third, cities create the conditions for more-democratic governance by bringing together people with common concerns. Living in close proximity with others, people need public schools, waste management, and sidewalks that are best provided collectively. They also face common threats: natural disasters or large public projects that disrupt neighborhoods or threaten places of historic value. It thus makes sense for people to organize into publics to obtain what they need to live well and protect themselves against harm. Banding together requires that they talk with each other and act in ways that maintain commitment and hold publics together. Under such conditions, democratic practices are likely to thrive.

Once governance mechanisms are in place to provide public goods and services and to regulate what can and cannot be done, however, they attract people who want to control them for other, more private and even personal reasons. Real estate interests act publicly to influence land use regulations so as to create development opportunities. Politicians use the local government, its many agencies, and subsidiaries such as public authorities to amass power or wealth. These governmental bodies have money to invest, supplies to purchase, contracts to assign, and services to buy, all of which can become favors to increase personal influence and lead to future rewards. Control of governments and governance more generally often become concentrated among a small segment of the business community, professional politicians, and other civic leaders. The financial resources and legal powers of municipal governments offer numerous opportunities to concentrate power amongst a few people with the consequence that the very democracy that cities foster is weakened. At these moments, oligarchy reigns and democracy is stifled.

Fourth, and lastly, cities encourage tolerance while simultaneously establishing the conditions for discrimination against and the marginalization of minorities, whether homosexuals, Mexican immigrants, African-Americans, or Muslims. As places where people of different ethnic and racial backgrounds, sexual orientations, religious beliefs, ages, and incomes are brought together, cities require tolerance to function. For the most part in the United States, tolerance exists toward diverse others (including strangers) with people mingling without overt animosity.[29] In fact, people who would otherwise be discriminated against frequently relocate to cities to escape the intolerance of their small towns. There, they experience an anonymity that allows them to embrace a lifestyle that suits them and to find others, others like themselves, with whom to mingle.

Of course, even within diverse cities people do not fully mingle in neighborhoods and workplaces. Numerous separations allow highly paid lawyers to occupy the same buildings as low-wage cleaning workers who do their jobs after the lawyers have left for the day, low-income households are unlikely to reside in middle-income areas, religious groups often cluster around their place of worship, and people with disposable income are more likely to be found at major sporting events than those who struggle from paycheck to paycheck. In a number of these situations, as we will discuss, such separations are more or less acceptable. They are not, however, innocent. Less influential groups are often marginalized or discriminated against in law offices, hospitals, and neighborhoods. Unable to wholly avoid each other, groups clash over religious festivals, ethnic parades, access to jobs, and the curricula of neighborhood schools. In worst-case scenarios, certain groups are ghettoized, fearful of going where they are unwelcome or likely to be physically assaulted.

These four contradictions intersect and their interconnections are numerous.[30] For example, economic wealth has political implications and those who command it often have better access to elected officials than the average citizen. Groups become politically dominant and are then able to pass laws banning legal protections for undocumented immigrants or convince the government to adopt green technologies. Elected officials channel government resources to encourage economic activities that serve already-wealthy workers and investors to the detriment of the less wealthy, thereby exacerbating inequality. Local groups defend a polluting industry against environmental activists because it provides employment.

The point is that cities are not simple machines producing a single product (for example, wealth) but complex and contradictory places in which a variety of diverse consequences are at play.[31] My goal with this book is to convey a particular understanding of how cities work, what they enable, and whom they serve. And while I have political biases that favor equality, sustainability, diversity, tolerance, and democratic governance, I am less interested in convincing readers to agree with me than in demonstrating the existence and consequences of these contradictory tendencies. To celebrate the city without acknowledging the deeply anchored nature of its antagonisms is to hide from reality.

Final Thoughts

Much of the urban literature clings to the point of view of a single scholarly discipline. If we view the core problem of human settlements as how people manage to live together, resolving conflicts and engaging in collective activities that have mutual benefits, then we are likely to cast the city as a political space and governance arrangements as central to its definition. If we are concerned with how people interact, identify with places, and form communities, then we will be more sociological and turn our attention to neighborhood dynamics, churches and bowling leagues, and the way people come together in public spaces. If our leanings are toward architecture and urban morphology, then we will approach the city in terms of its buildings, sidewalks, road networks, utility systems, and playgrounds and emphasize the visual. This reduction of the city to that which can be viewed through a single and narrow (disciplinary) lens, however, is not what I wish to do here. This is not a book about the culture of the city or its economy, but rather aims to cut across these perspectives. The reader might feel compelled to situate the four contradictions in disciplinary categories, but this is not my intent. As a whole, the book is meant to blur such boundaries.

I am not, however, claiming to be inter- or even multidisciplinary. To do so verges on having no point of view whatsoever. My aim is neither to encompass all disciplinary approaches nor to be indiscriminate as regards those aspects of the city that matter. Moreover, while I draw on a strong belief in the importance of economic relations for both how people live in cities and whether cities thrive or not, the book does not end with "the economic" all-powerful and determining. Any reduction of the city to its economy relegates the rest of the world to epiphenomena. Cities are highly complex formations of humans, technologies, built forms, and nature and I use this understanding to move closer to the concerns of people who live and work, visit and invest there. Their lives are not lived in disciplinary categories.

My objective is to identify what matters to those who use the city. This means not merely describing what happens in as unfiltered a way as possible. We must also recognize that the choices we make in the face of the perplexing consequences of these contradictions are what give the city—and any particular city—its character and what implicates us in the moral ties that define neighborhoods, church congregations, sports clubs, and the city itself. Because they emerge out of contradictions that challenge shared values, these are not simply physical conditions. They also embody what we as a people deem acceptable in living together. They have moral implications.

Much, if not the great bulk of, contemporary popular commentary, scholarly research, and public policy focuses on the city as an economic phenomenon.[32] For that reason, and keeping in mind the caveat above, I begin with the contradictory presence of wealth and poverty. The opportunities for people to earn money and spend their incomes have a significant impact on whether a city's economy grows or declines. The ways in which people live in cities (and the workings of its economy) depend, as importantly, on how people treat the natural environment, how the city is governed, and whether tolerance or intolerance reigns. Heavily influenced by the form and functions of a city's economy, these other consequences are nevertheless not determined by it. We turn, now, to this first contradiction: the ways in which cities enable the creation of wealth and block its fair distribution. In too many cities, the consequence is great affluence side-by-side with unshakable poverty.

TWO

Wealth, Poverty

Driving along the streets from one side of Phoenix or St. Paul to the other offers visual confirmation to the traveler not only that people live differently but that the quality of the places in which they live is alarmingly dissimilar. Some people occupy neighborhoods with trash-strewn lots and cracked sidewalks; their homes are dilapidated and the local retail area pockmarked with vacancies. Others live near upscale shops and well-manicured parks; their homes are large and inviting. The co-presence of affluence and poverty is there for all to see.

What one witnesses on such a journey is a playing out of a contradiction between two mutually antagonistic but inseparable forces. On the one hand are the forces that generate and concentrate wealth; on the other hand, those that create and concentrate poverty. Resources and opportunities are gathered into the hands of a few people, while others are denied them. Lacking strong mechanisms to redistribute wealth, poverty and inequality are the result. Not just the result, they are seemingly necessary if great wealth is to be enjoyed by a small minority of the city's residents. And while cities do not create this contradiction, they provide the conditions that enable it to flourish.

The purpose of this chapter is to describe and document the relationship that exists between the forces that generate wealth and those that generate poverty. We begin with a quick immersion into what it means to live amid these antagonistic forces and then move to the generation of wealth, the concentration of wealth and poverty, and the interdependence of the two. Emphasized throughout is the inseparability of public wealth and private outcomes.

Living amid Wealth and Poverty

Because of cities, great wealth is possible. Unlike small towns and rural areas, cities create large and diverse markets and open numerous avenues to financial profit. They bring together diamond merchants, textile firms, drug stores, and accountancies that compete against each other while spurring innovations in product quality, service, and marketing. Out of competition and cooperation emerge novel commodities that fuel consumer demand and expand commerce. Banks provide financing for new ventures, law firms offer legal services, insurance companies protect against risk, and business consultants give advice. Within commuting distance are workers with a variety of skills and education with publicly financed transportation networks connecting homes to places of employment. In brief, a multitude of opportunities exist to distribute profits to business owners and investors and jobs to residents and commuters.[1]

Cities also support large (even international) airports, convention centers that draw visitors from around the country, clusters of colleges and universities and hospitals, and extensive networks of public parks. The wealth of cities is not just private (that is, due to and in the hands of individuals and corporations), it is also public. To ease commerce and make life not just bearable but pleasant for its residents, governments build roads and subways, operate schools, set aside land for parks and playgrounds, attend to public health with child vaccination programs and food sanitation regulations, and use their police powers to assure public safety. They invest in museums and professional sports stadiums and support hospitals, churches, and private educational institutions by reducing property tax burdens and providing fire protection, sewage services, and traffic control. Governments also subsidize urban redevelopment projects, regulate air and water quality, and build and maintain public parks. All of this is part of the wealth of the city and essential to what makes it attractive to households and investors.

Public wealth thus enhances the creation of private wealth. As illustration, local schools prepare children to enroll in college, subsequently earn graduate business degrees, and take high-paying positions in major corporations. Governments finance waterfront development and the resultant projects boost adjacent property values; they extend light-rail systems to bring more people downtown to shop and work. By comparison to private wealth, public wealth is more widely shared. Even poor neighborhoods often have bus access, an elementary school, recreational facilities, and emergency health care.[2] Although affluent neighborhoods and poverty-stricken ones frequently coexist in the same city, the blame for this disparity almost wholly lies with the distribution of private wealth and not with the distribution of public wealth.

Adept at generating private wealth, however, cities fail at distributing it in an equitable manner.[3] Some businesses—banks, investment firms—become quite large and their executives extremely well paid with yearly salaries multiples above what are earned by the clerical workers, technicians, and low-level managers that they employ. A few landlords and investment companies own a disproportionate share of the city's large apartment and office buildings. Universities, hospitals, world-class museums, and foundations pay their chief executives and most-skilled staff amounts that greatly exceed that earned by the average store clerk, public bus driver, or electrician. Having incomes well beyond subsistence needs, these people have surplus funds to be invested in real estate, stocks, and bonds to generate even greater wealth. Money, as the political philosopher Michael Walzer reminds us, is a primary good that can be used to buy other goods—education, health care, political connections, prosperous neighborhoods.[4] With money comes influence and with influence money. People neither start equal in wealth nor end up there. By fostering the concentration of wealth among a small number of highly skilled workers, businesses, property owners, and entrepreneurs, cities exacerbate inequalities.

A few people become very wealthy and many households live well, but a large portion of the city's population has barely enough money to manage a precarious existence. The financial distance between those at the top and those at the bottom is obvious to anyone who travels through a major city in the United States. Multi-million dollar homes occupy some neighborhoods and low-rent and run-down apartments others. The affluent purchase courtside seats to watch professional basketball and the homeless line up outside a charitable shelter. Restaurants in the most expensive parts of the city offer meals costing hundreds of dollars while fast-food franchises, a cheaper but less healthy alternative, crowd into poor neighborhoods. Wealthier areas have good public schools or their children attend private academies, while in poor neighborhoods children receive a substandard education. The only conclusion, it seems, at least for US cities, is that private wealth more than public wealth matters for living well.[5]

This interplay of private and public wealth and their distribution across the residents of the city is the central concern of this chapter. Although cities are great at generating wealth, they perform poorly in distributing it such that all who reside there can prosper. To make this argument, we begin with how cities foster the conditions for wealth-generation and concentrate it among a few businesses, institutions, and households. Why do some people disproportionately benefit? How does poverty become entrenched? And, why do wealth and poverty occur together and persist?

Generation of Wealth: The Public Contribution

To understand how wealth is generated, we need, first, to recognize the extent to which the prosperity of private individuals and organizations depends on the presence of public wealth; that is, on the good roads, libraries, parks, and public museums that enable commerce, make cities attractive to households and businesses, and mitigate the excesses of private sector activities. Once done, we can turn to the creation of private wealth. Our focus throughout will be on the geographic concentration of workers and capital and the way this fosters growth.

Until public wealth is put in place, cities do not even begin to generate private wealth.[6] Without roads and wharves, protection of private property, a supply of potable water, banking regulations, and laws regarding sanitation, among the many contributions of government, people would not be able to live in dense and large settlements, ship goods to market, and amass fortunes. During European colonization, local elites often taxed property owners so that warehouses and market stalls could be built, roads made passable, and lands (usually what became the village green) opened for grazing sheep and cows. Internal improvements such as bridges and canals were also key to urban development. In the late nineteenth century, the City of Houston dredged a ship channel to the Gulf of Mexico that greatly increased commerce and, in the decades just after the Second World War, numerous local governments subsidized and coordinated urban redevelopment projects to attract private investment, boost property values, and generate jobs. Using government policies and private sector financing, cities like San Diego have launched initiatives to attract technology firms and health laboratories. More recently, a number of cities have built municipal broadband networks that provide residents with free access to the internet. The increasing importance of the internet to daily life has convinced these local governments that such services deserve to be publicly available.[7]

Throughout the history of American cities, the creation of private wealth has depended on government regulation of interest rates, zoning laws that protect private property from unwanted encroachments, the granting of rail rights-of-way, and suppression of labor unrest. Today, the wealth of cities is widely believed to be a direct consequence of the presence of educated labor and the public and private universities that support research, generate innovations, and train individuals to start businesses.[8] Equally significant are the global airport connections, the policing of crime, and governmental regulations that protect real property and investments.[9]

Local governments also construct and maintain buildings and infrastructure that enrich the daily lives of residents, facilitate commerce, and give texture to the urban landscape. Spread across the city, fire stations house emergency vehicles, parks provide a respite from the city's hard surfaces, and libraries offer learning opportunities to immigrants. Public streets enable bakery vans to deliver bread to the city's restaurants each morning. Consider Long Beach, California, a city of approximately 462,000 people in 2010. At that time, its city government managed 132 public buildings with a replacement value of $800 million. (See table 2.1.) In addition, it was responsible for local and arterial streets, sidewalks, curbs, alleys, catch-basins (3,000 of them), and storm drains, pipes, and channels. These nondescript public assets are frequently overshadowed by the large-scale infrastructure that cities build: Denver's $4.8 billion international airport that opened in 1995, Portland, Oregon's light rail system with its 52 miles of track and 87 stations, and Miami's Metromover (monorail) with 21 stations and a daily ridership of 35,000 passengers, to name just three. Cities grow because of these buildings and structures and because of the regulations, laws, and services that governments provide. Most importantly, they thrive when these public contributions are of high quality, are appropriate for a city's size, and attend to the city's diverse people and functions.

Over the centuries, public wealth has been generated in a number of ways. Until the formation of local governments, trading companies made the improvements (as in New Amsterdam in the early seventeenth century), local elites pooled their resources, or taxes were levied on property owners. Land heretofore unoccupied or used by Native Ameri-

Table 2.1 Buildings and Structures Owned and Managed by the Government of Long Beach, California, 2007

Buildings	Number
Park buildings	58
Fire stations	30
Police stations	17
Libraries	13
Health facilities	4
Other	10
Total	132
Structures	**Miles**
Curbs	1,500
Local streets	556
Arterial streets	259
Alleys	221
Storm drains, pipes, channels	180
Sidewalks	163

Source: City of Long Beach, "A City in Need of Capital Investment" (Long Beach, CA: Public Works Department, 2007).

cans was forcefully or deceptively appropriated and rights-of-way established. Until the late nineteenth century, many internal improvements were small and ad hoc rather than large and systematically provided. One exception was the Erie Canal that began operation in 1825 and connected New York City to an interior rich in natural resources and agricultural products. Not until laws were established for the election or appointment of local officials and mechanisms devised for collecting taxes from property owners were local governments able to offer infrastructure and public services such as policing or fire protection on a citywide scale.[10] Even then, the development of public wealth depended to a great extent on the creation of new technologies such as water pumping stations, gas lighting, and traffic signals along with the training of engineers, architects, and accountants who could manage public funds, design and build public projects, and craft and enforce public health, construction, fire safety, and land use regulations.[11]

Public wealth, of course, is no more naturally occurring than private wealth. It depends on the disposition of elected leaders (what is usually termed public policy) and entails the willingness of governments to tax private wealth and designate funds to build roads and community centers. Public wealth also depends on the extent to which economic elites recognize the benefits of public goods and services. Never far from decisions about expanding libraries or extending the local public bus service are concerns about the proper role of government in relation to private economic activity and individual freedoms. Do government regulations enhance or diminish economic growth? Are taxes a burden or an opportunity? Should the local government have more powers or fewer? What is the proper relation of the municipal to state and federal governments? These questions, and the debates that they engender, linger beneath all decisions about public wealth, whether it is to provide more day care facilities or sell the local airport to a private firm.[12]

The Generation of Private Wealth

As for private wealth, the pathways to its creation are innumerable and cities (in part through their governments) support these pathways in a variety of ways: facilitating the concentration of consumer and labor markets, fostering a division of labor that enhances productivity and innovation, increasing the scale and density of activities, simplifying access to capital, and engendering the agglomeration that gives rise to a multitude of opportunities for sharing ideas, resources, and space.[13]

Consider, first, the concentration of economic activity. Brought together in high-density settlements, households form local markets that enable providers of goods and services to grow in size. These markets give rise to banks that provide loans for the purchase of homes and the operation of businesses. At the same time, numerous individuals are available to be hired as workers such that businesses can grow and thrive. In short, large markets set the base conditions for individuals to expand economic activity and make money. Cities "are places where good ideas get a hearing from investors, where entrepreneurs can rapidly prototype and then test-market sophisticated products, [and] where a start-up firm can draw on a thick talent pool."[14]

In villages and towns, drug stores are small, restaurants are few (often a single diner for the locals), and hardware stores nonexistent. Shoe and clothing stores are unlikely to be nearby and although a lawyer, an accountant, or a doctor (usually a general practitioner) might set up business, more specialized professional services can only be found elsewhere. When the town grows in population and becomes a city, businesses can be bigger, more numerous (as with automobile dealerships), and more diverse. A city has a large number of people who need food, clothing, and shelter along with medical services, legal advice, entertainment, and appliances. Dallas has multiple stores selling musical instruments along with hospitals that have the latest medical technologies; the sparsely populated Archer City, approximately 150 miles away, offers neither. And as businesses emerge and the local government expands to service a greater diversity of needs, they augment the demand for accountants, office supplies, and computer repair services.[15]

Individuals who start businesses to serve this demand can decide either to remain small, targeting only a portion of the market or confining themselves to one area of the city, or grow to serve an expanding population. Those who take the latter path have the option of becoming quite big, leading to control over a large share of the market for, say, lumber or newspapers. One or two firms become the major suppliers of concrete for construction with only a handful of companies building major office complexes and sports facilities. A clothier expands his range of goods, builds a larger space, advertises daily in the local newspaper and on the radio, and becomes the city's major department store. Fast-food chains, clothing and food franchises, and corporate pharmacies establish multiple locations across the city and metropolitan area. Consumers travel downtown from their neighborhoods to shop at the city's largest furniture store, while heating oil firms can easily access multiple clients. Not only is demand relatively concentrated, but roadways, subways, and public buses enable purchasers easily to access goods and services and businesses inexpensively to deliver to customers. A large market, close at hand, enables all of this to happen.

The combination of transportation accessibility and the multitude of individuals and businesses in need of supplies and advice drives businesses to grow in market share, expand in size, and achieve significant sales. Business firms in cities are generally larger than they are in non-metropolitan areas. Whereas in 2011 only 0.3 percent of all firms in the country had more than 500 employees, the percentage in metropolitan areas was much higher: 6.4 percent in Tucson (AZ), 8.0 percent in Durham (NC), and 4.8 percent in Oklahoma City (OK).[16] These differences also hold for smaller firms that have between 20 and 500 employees. Bigger and more prosperous businesses, moreover, have revenues that can be translated into larger profits for the owners ostensibly compensating them for their business acumen, the financial investment they have made, and their management skills. In addition, the greater the market share and the fewer the competitors for advertising services or meat provision, the more leeway business owners have in setting prices. The existence of big consulting firms, automobile-makers, and telecommunication companies that serve regional, national, and even global markets does not negate the importance of cities to their beginnings or to their continued operation. Cities put forward the concentrated markets, the public infrastructures, and the regulatory technologies that allow businesses to deliver their product at a scale that enhances profitability and leads to earnings that finance expansion beyond a city's boundaries.

Key to the emergence of a large and diverse array of firms is the concentration and availability of labor—more specifically, of a wide range of skilled and unskilled workers willing to be employed at various hourly wages and salaries. Many people migrate to cities because they wish to find jobs in advertising agencies or financial service firms; be near hospitals where they can open a medical practice; own a pizza joint; play in the local symphonic orchestra; or work in the city's schools. Others are pulled less by occupational interests than by the allure of the city as a place of opportunity. There, they might spend their days or nights as waitresses, cab drivers, garment workers, sales clerks, and security personnel. Most working-age adults are employed by others rather than being self-employed. They are attracted by the higher salaries in cities. (See table 2.2.) In Miami, the average yearly salary in 2014 for a chef was $57,000; in the non-metropolitan Butler County in Missouri, it was $21,000. A firefighter could earn $71,000 in Miami and $29,000 in Butler County. For those starting businesses or nonprofit organizations, and for governments, the city offers vast numbers of potential workers with a wide range of skills and experience. Without those workers, businesses would remain one-person firms or family operations with meager prospects for significant growth.

The extensive division of labor makes cities even more valuable for the creation of private wealth. As more and more people come together, opportunities exist for some to specialize in violin repair or financial

Table 2.2 Annual Mean Wages (US$) by Occupation for Selected Large Metropolitan and Non-metropolitan Areas, 2014

	Construction Manager	Firefighter	Chef	Clergy	Security Guard
Large metropolis					
Miami-Fort Lauderdale	94,830	71,390	56,830	52,100	22,850
Des Moines	80,070	35,080	37,880	45,820	36,710
Non-metropolitan area					
Bailey County (TX)	78,430	39,840	38,590	42,170	23,290
Butler County (MO)	73,810	29,230	20,920	38,270	26,900

Source: Bureau of Labor Statistics, "Metropolitan and Non-metropolitan Area Occupational Employment and Wage Estimates," posted 2014, accessed April 3, 2015, www.bls.gov/oes/current/oes2900003.html.

wealth management. People gravitate to specific jobs and, as their incomes increase, desires expand and people want a greater variety of goods and services. This further pushes businesses into specialized markets and people into different occupations. With specialization, workers and businesses become better at what they do and productivity increases. Profits rise and are distributed, albeit disproportionately, between workers and owners. Here is another way—heightened productivity through a division of labor—in which cities enable wealth to be generated.

Wealth is amassed not only by those who own businesses. Many people relocate to cities for fame, not just for fortune. They might hope to become a television newscaster, a rock star, CEO of an internet startup, a famous chef, or a widely published author. Or, they might join a hedge fund firm, become a department head at a major hospital, or seek employment as a statistician with a professional sports team. Workers in many of these occupations and industries also have the potential to become quite prosperous. Large corporations, banks and financial service firms, universities, hospitals, philanthropic foundations, and real estate development companies (among others) hire highly educated people and pay them significant salaries. Health clinic doctors, professors at major research universities, and heads of public authorities all earn sums of money greater than what they need to live comfortably.

As illustration, see the salaries of university professors as shown in table 2.3. Those working in major cities earned an average yearly salary in 2013–2014 well above the average incomes of the respective metropolitan areas. At colleges and universities outside the major cities, the salaries are much lower. A good portion of this difference can be attributed to the larger size of the former universities and their affiliation with medical schools, but much also has to do with the synergies of the city that generate higher (and high) salaries in the businesses and organizations there.

The concentration of markets for goods, services, and labor generates wealth in other ways as well; it enables economies of scale and creates densities that facilitate commerce. Economies of scale are the correlate to the productivity gains from a division of labor; they occur as a consequence of the increasing size of things and activities. The basic idea is that an activity has a certain scale at which it is efficient and other scales at which it is not. A theater company might be profitable putting on plays for an average-size audience of 300 patrons over the run of the play, but not for 225 people. At 300 theatergoers, all costs are covered and after that, as the audience increases, the profits continue

Table 2.3 **Annual Professorial Salaries (US$) in Large versus Small Cities, 2013–2014**

	Full Professor	Assistant Professor
Large cities		
Chicago: University of Chicago	210,700	105,600
New York: New York University	195,700	110,100
Los Angeles: California Institute of Technology	182,100	116,200
Boston: Boston University	161,600	93,200
Small cities		
Williamstown, MA: Williams College	140,000	78,200
Muncie, IN: Ball State University	90,000	58,700
Joplin, MO: Missouri Southern State	75,300	47,600
Rocky Mount, NC: North Carolina Wesleyan College	52,400	46,400

Note: California Institute of Technology is in Pasadena in the Los Angeles metropolitan area. The median household incomes in 2013 in Chicago, New York, Boston, and Los Angeles respectively were $60,564, $67,786, $72,907, and $58, 869. See Amanda Nuss, "Household Income: 2-13" (Washington, DC: US Bureau of the Census, 2014).
Source: 2013–2014 AAUP Faculty Salary Survey, accessed April 24, 2015, http://chronicle.com/article/2013-14-AAUP-Faculty-Salary/145679#id=table.

to grow, limited only by the size of the venue and the ticket prices people are willing to pay. A similar phenomenon occurs with elementary schools, department stores, cardiology practices, and policing. This is one of the reasons an architect might find it difficult to establish a profitable practice in a rural area; there are simply too few clients to make a decent living. At a certain size, activities become more profitable and efficient. This is true even for such governmental activities as building code enforcement that need a certain number of buildings to keep inspectors busy. These economies are not fixed but rather vary by the size and affluence of the city, prevailing technologies for producing goods and delivering services, and the accessibility of supportive public infrastructure such as mass transit. Outside of cities, these conditions are less likely to be present.

More than just the number of buyers of goods and users of services matters. The greater the geographical density and accessibility of consumers and users of public services, the lower the costs—transaction costs in economic parlance—of matching them to demand. The costs in time for a family to travel to an optometrist or a playground, for a furniture store truck to deliver a dining room table and chairs, or for a shelving installer to drive to a client's store are—depending on traffic congestion—less than in lower density places. High density and the proximity that it creates smooth the interaction of businesses as those businesses sign contracts with each other, ship raw materials and finished goods back and forth, and share in the pool of workers. A company that assembles custom-designed windows will have access to metal manufacturers, glass supply houses, makers of rubber gaskets, and hardware manufacturers. In a relatively large and dense market, numerous opportunities exist to organize businesses and deliver public services efficiently and, for private enterprises, profitably. Size and density, coupled with public transit and public streets, enable scale economies to be realized, thereby enhancing the creation of both private and public wealth.

With their size, density, and diversity of economic activities, cities also make capital more accessible. Entrepreneurs require money for investment in new businesses and existing businesses need operating capital so that they can purchase raw materials, hire consultants, and pay their workers while themselves waiting to be paid for the products that they have sold. Businesses generate cash and that cash has to be put someplace safe until it is needed. Banks provide this service and then lend the money to other businesses as well as to consumers wishing to purchase homes or put an addition on their house. Banks thrive where numerous businesses are involved in a multitude of financial transactions. Money can be made by facilitating the flow of funds from businesses to workers, from businesses to suppliers and subcontractors, and from businesses to government and into investments. Of the top 25 banks by total assets in 2014, all but three were headquartered in a major city with over one-half of them in New York City, Chicago, and San Francisco.[17]

It is not just banks that locate in cities to take advantage of the money circulating in their economies. Cities also attract and give rise to investment firms, venture capital funds, and wealth management operations which, in turn, attract accounting firms, law offices, developers to build office buildings to house all of them, and janitorial services to clean offices at night. And, even though banking and financial services easily cross city and even national boundaries, they seem to be most profitable when clustered in cities. There, they add to private wealth.[18]

One illustration of the pull of cities is the concentration of major corporations. Each year, *Fortune* magazine ranks the 500 largest closely-held and public corporations in the United States by gross revenues.[19] (Public corporations are those with stocks that are traded publicly rather than being privately owned and untraded.) In 2014, 27 of the top 50 corporations had their headquarters in large and medium-size cities that were the centers of their metropolitan areas. These were places like Houston, Charlotte (NC), and New York City. Another 19 were outside the central city but still within a major metropolis. Examples are Microsoft (ranked 34th) in Redmond (WA) in the Seattle metropolitan area and PepsiCo (ranked 43rd) located in Purchase (NY) within New York's metropolitan region. Only four of the top corporations were in small towns or in small metropolitan areas. Wal-Mart (the largest of these corporations by gross revenue) is headquartered in Bentonville (AR), Marathon Petroleum in Findlay (OH), State Farm Insurance in Bloomington (IN), and Dow Chemical in Midland (MI).

The spatial concentration of activities, its agglomeration, is credited not only with providing large and accessible markets, lowering transaction costs, and increasing the potential for economies of scale, but also with providing innovations that enable urban economies to adapt and change.[20] The key here is the large size and diversity of the economy. The most important factor is the density of human interactions. The argument is that where large numbers of businesses and individuals exist close together, many more interactions are likely to occur among them. This engenders heightened competition among firms, thereby encouraging them to find ways to maintain and enhance their market share and accelerating the flow of ideas. Known as knowledge spillovers, these ideas are often adopted by others and used to develop new products, manufacturing processes, technologies, and services. The resultant innovations then enable the local economy to expand and replace goods and services no longer in demand or currently being imported.[21]

Consider Detroit and its automotive industry. That industry emerged from the metal-working milieu that existed there in the early twentieth century. In New York City, the synergies among advertising, magazine and book publishing, and fashion have created a robust media concentration. Toledo (OH) was able to develop a solar panel industry because of the knowledge and skills developed in an earlier era of glass manufacturing. As a last example, Boston has replaced manufacturing with financial services, expanded its medical sector, and nurtured its universities by drawing on an educated population with a diverse set of skills.[22] These stories of economic success are a function of the agglomeration qualities of cities and the knowledge spillovers and mutual learning that flow from them.

Agglomeration not only creates wealth through innovation and expansion of the economy, it also does so by increasing densities and boosting land values. For many people and businesses, much of their wealth resides in the real property that they own and the land and buildings that they use. Cities enable this wealth by making specific locations highly desirable and thus causing those who want to be there to pay a good deal of money to do so. Land is made to be scarce and its value is reflected in the value of buildings. In 2013, the median value of owner-occupied housing units in the United States was $176,700. If you lived in San Jose (CA), however, the value was quite different—$560,400—while in Stamford (CT) it was $515,400 and in Los Angeles $446,100.[23] This is not, moreover, a matter of a value inherent to a location or solely of the economic competition for land, but also of the many activities of government that make some places more desirable and valuable than others.

Public investments are major contributors to the benefits of agglomeration. As Alex Marshall has noted: "Creating places is almost wholly a product of public, political, and tax-payer financed decisions."[24] When governments build highways and bridges and allow bus and commuter rail agencies to construct access roads, tracks, and terminals in the city center, these cities become highly accessible and thus desirable as places to locate office buildings, department stores, and high-end retailing. Most retail businesses want to be easily reached and large insurance offices and banks want to minimize the cost of commuting for their workers. Big entertainment venues such as clubs or arenas attract large crowds by locating in high-density cities where roads and transit lines converge. Add to this the many individuals who work in these firms and wish to be close to their place of employment with easy access to grocery stores, movie theaters, hairdressers, and restaurants. Governments make cities even more attractive and valuable by protecting property values through zoning and other land use regulations, building parks, keeping the rivers clean and the air unpolluted, and stipulating property boundaries and recording real estate transfers. As important, they provide political support and financial subsidies to local elites that strengthen certain industries (for example, biotechnology) and make the city more attractive to tourists and conventioneers.[25]

Public infrastructure also generates wealth by adding value to location. The resultant proximity to services and amenities drives up the price of land in the center or around various nodes where transit lines converge and densities have peaked. Property companies emerge to buy up and manage apartment buildings and office space, while households purchase expensive apartments near restaurants, entertainment, and the office buildings in which they work. This value is captured through private ownership. And, as the city grows in population and economic activity, land becomes even more valuable. The wealth of owners increases as the land appreciates.[26]

All of this happens because of roads, bridges, and subway lines, water and sewer systems, publicly regulated electrical utilities, schools systems, public health regulations, property use restrictions, laws protecting private property, and fire and police services.[27] These public services enable commerce, support economies of scale, encourage density and agglomeration, and create the conditions for people and firms to amass wealth. Cities that grow, moreover, build public wealth. Their governments add water treatment facilities, high schools, parks, playgrounds, traffic lights, and environmental regulations. These assets have both an intrinsic value and a value that extends outward to contribute to the generation of private wealth. A neighborhood park provides residents with opportunities to socialize and enhances adjacent property values. The central bus terminal enables patrons to transfer from one line to the next and brings consumers into the city to shop. Regulations—another form of public wealth—provide the legal foundation on which private wealth is built and create public spaces (such as parks) without which the city and its residents would live a less rewarding existence. Absent these public services, a city would neither grow in size nor reach the density necessary for sustained economic growth.

Private wealth, of course, is not simply dependent on public wealth; the reverse is also true. Local governments are able to build infrastructure, inspect housing for code violations, and provide day care to the extent to which they can extract revenues from households and economic activity via business, sales, and property taxes. A number of cities levy taxes on resident income and charge fees to developers wanting to build luxury apartments or office buildings. The more activity and wealth, the more tax revenues can be captured and the more money is available to run programs and pay off the bonds used to construct schools, bridges, and recreation centers. The synergies are important if the city's economy is to grow and the city is to remain attractive. This does not mean that public wealth follows inexorably and in proportion to private wealth. To the extent that taxes diminish business profits and the disposable incomes of households, taxpayers (and corporations in particular) resist even as they benefit from them. How do public investments, they ask, serve local business and property owners, the interests of investors, and those (for example, poor children) in need of public services such as free school lunch programs?

What will it cost in taxes to provide these supports? Here is where law and politics merge as local officials, economic and cultural elites, and community advocates debate what taxes should be levied on whom, what capital investments should be made, and which programs should be expanded, contracted, eliminated, or launched.[28]

Certain forms of public wealth are privately owned. Think of nonprofit museums such as the Carnegie Museum of Art in Pittsburgh or the Getty in Los Angeles, symphonic halls in Denver and Fort Worth, and private universities such as Rice University in Houston, and the Rhode Island School of Design in Providence. Then there are the hospitals and professional sports arenas that one finds in almost all large cities. These are public assets: they are open to the public, contribute to the urbanity of the city, and figure into the city's economy. The public acknowledges this by having the government subsidize their construction and renovation. Local governments recognize the need to protect and enhance these assets.[29] In 1993, for example, voters in the Pittsburgh metropolitan area approved the formation of the Allegheny County Regional Asset District that dedicates county-wide tax proceeds for the support of cultural institutions, libraries, parks, and sports facilities, most of which are in the city of Pittsburgh.

To such public wealth, I would add private foundations that contribute to the quality of life in cities by making them livable and attractive. Cleveland is a better city because of the presence of the Gund Foundation and the Cleveland Foundation. A similar claim can be made for Seattle with the Bill and Melinda Gates Foundation, Baltimore with its Annie Casey Foundation, and Chicago with the MacArthur Foundation.

Another form of private wealth that has public value can be found in buildings. Among a city's cultural assets are public buildings such as city halls and libraries as well as private buildings such as residences, famous office towers, and factories. The Colonial-era homes in Society Hill in Philadelphia and on Beacon Hill in Boston contribute to the historic and public character of these cities—and to their uniqueness. The early twentieth-century office buildings in Chicago (such as the Monadnock and Wrigley buildings) and, more recently, the John Hancock Center, the Willis (formerly Sears) Tower, and the Aqua apartment building by Studio Gang Architects are all valued contributions to the city's skyline and streetscape. The casinos of Las Vegas, the Seattle Space Needle, the Arch in St. Louis, and the Darwin Martin House in Buffalo designed by the famous architect Frank Lloyd Wright provide character and cultural significance. Together, these buildings and structures make a city desirable, give it identity, and enrich the urban experience. In doing so, they expand the city's wealth.[30]

If a person wants to become wealthy, she must (with exceptions) go to a city. If she wants to become exceptionally wealthy, the city is unavoidable. There, numerous conditions nurture the creation of private wealth. These conditions, however, operate robustly only in the presence of a strong public realm.

Concentration of Wealth

Cities contribute to the generation of wealth and enable it to become concentrated in the hands of a few individuals, families, and businesses. To this extent, they are places where inequalities in wealth and income, along with stark variations in the quality of peoples' lives, become visible.

Consider a few examples. Each weekday morning, senior corporate executives are picked up for work by a limousine service, while a retail clerk, forced by high rents to live far from his job, walks blocks to a public bus stop in order to make his way to the nearest light-rail connection. When summer comes, the executive flies her family on a private jet to a vacation resort for an extended stay or to their house in southern France; the retail clerk drives the family to the beach for the weekend. Uniformed doormen hail taxis for the residents of high-end apartment buildings, while homeless individuals push their shopping carts (loaded with their possessions) along the sidewalk behind the bus terminal; shops sell expensive Italian clothing for men while corner liquor stores in distressed neighborhoods sell beer and alcohol from behind protective plexiglass barriers.

Documenting the concentration of private wealth, however, is not easy, in part because the very wealthy often have multiple residences—not only a place in San Antonio where they have their business but a vacation home in Breckenridge, Colorado, that they visit throughout the year. Although they might have a primary residence in one city, their wealth is geographically dispersed across multiple properties and across investments that move incessantly through the global circuits of capital. For the rich, only a small proportion of their wealth is spatially fixed. In 2010, for example, the wealth of home-owning US households in the lowest income quartile was $282,000, while that for home-owning households in the highest income quartile was $1,572,000. For the former, 36 percent of that wealth came from home equity compared with 16 percent for the latter, despite the fact that households in the highest income quartile are likely to own multiple properties.[31] These data do not establish the urban concentration of wealth. The issue for my argument, though, is one of trying to understand how cities concentrate wealth in individuals and families, not how wealth becomes situated in cities.

One way to measure this concentration is by using earned income. Census data show that most high-income households live in highly populated areas. Of the households earning at least $191,000 per year between 2006 and 2011 (that is, the top five percent of the national income distribution), 52 percent of them lived in the 50 largest metropolitan areas.[32] The five metropolitan areas with the largest percentage of these households were San Jose–Santa Clara ("Silicon Valley"), Washington, DC, San Francisco, New York, and Boston-Cambridge. Silicon Valley led the way with nearly 16 percent of its households in the top 5 percent of the income distribution. Boston-Cambridge was the lowest of the group with slightly less than 10 percent of its households this wealthy. If one considers all metropolitan areas, not just the most populous ones, Bridgeport-Stamford, Connecticut, led the way with almost 18 percent of its households in the top tier of income.[33] Such concentration is to be expected given the high wages and salaries paid in large cities.

High incomes, however, are not evenly shared across a city's labor market. Income inequalities are quite striking. Of the 50 largest cities in the country, Atlanta leads the way with the richest households earning 19 times what the poorest households earn.[34] San Francisco is not far behind. Even in the city with the lowest ratio—Virginia Beach, the richest households earn 6 times what the poorest households earn. (See table 2.4.) A study of the income gap for New York City in 2013 found that the city's top 5 percent of households earned 88 times what the lowest 20 percent earned.[35] In a city where the median household income was $52,000 and 1.7 million people had incomes below the federal poverty threshold of $23,000 for a family of four, the top households took home that year a salary of $864,000. This unequal distribution of income is highly correlated with the unequal distribution of wealth. An individual with less than $40,000 a year in income in 2010 had only, on average, $3,000 in wealth, whereas those earning over $110,000 per year had net assets of $200,000.[36] The poorer you are, the less wealth you have and the less wealth you have, the less able you are to weather unexpected events such as a health issue or sudden unemployment.

Table 2.4 Inequality in Selected Cities of the United States, 2012

City	Inequality Ratio
Greatest inequality	
Atlanta	18.8
San Francisco	16.6
Miami	15.7
Boston	15.3
Washington, DC	13.3
Least inequality	
Virginia Beach	6.0
Arlington, TX	7.3
Mesa, AZ	7.5
Las Vegas	7.7
Wichita	7.7
Average of fifty largest cities	10.8

Note: The Inequality Ratio is the ratio of household income in the 20th percentile to household income in the 95th percentile.
Source: Alan Berube, "All Cities Are Not Created Equal," The Brookings Institution, posted 2014, accessed November 11, 2014, www.brookings.edu/research/papers/2-14/02/cities-unequal-berube.

How do wealth and incomes become concentrated among relatively few households and businesses? The answer begins with the observation that cities exacerbate advantages and disadvantages. This has consequences for who has access to and benefits from the private and public wealth that they generate. The answer does not stop there, however. We must also attend to the effects of place and the ways in which cities concentrate opportunities for advancement, the value of social connections, and the influence of government.

First, and central to any increase in inequalities, is the existence of initial advantages such that people with greater capabilities—education, access to resources (for example, inherited property, political connections), and family backgrounds and cultural experiences—are better positioned to profit from opportunities they generate or that come their way.[37] Those lacking these advantages are less able to amass wealth.

The computer revolution of the late twentieth and early twenty-first centuries led to such success stories as Bill Gates, who built the software giant Microsoft, and Jeff Bezos of the online retailer Amazon. Both started with advantages. They came from middle-class families, attended good schools (Gates at Harvard University and Bezos at Princeton University), had access to existing computer technologies (in the case of Gates) and financing (in the case of Bezos, who initially worked on Wall Street), and were blessed with entrepreneurial personalities.[38] As they developed their ideas, they also embedded themselves in social and financial networks that provided capital to start and expand their businesses. Once they became successful, resources and talent flowed to them.

Cities—certain cities—are also an advantage. It matters for upward mobility where one grows up. Cities like San Jose, Salt Lake City, Seattle, and Minneapolis are much more favorable environments than Fayetteville (NC), Detroit, or New Orleans. Children in the latter cities live with high levels of residential segregation and income inequality, poor schools, and weak social networks. Consequently, they are less likely to improve their circumstances than those who grow up in more advantageous places. Lacking upward mobility, they are also likely to live the remainder of their lives there. Certainly, many struggling cities (for example, Philadelphia, Buffalo) have pockets of affluence with the children lucky enough to live in these areas often isolated from the city's disadvantages. Their prospects are not so bleak.[39]

Central to these geographical effects is the ecology of neighborhoods. The value of anyone's home is dependent on the cost to build the house, its location, and the quality of the surrounding homes. When assigning a value to a house or apartment building, real estate appraisers, banks, and other intermediaries in the housing market (such as mortgage brokers) take adjacent properties (and thus location) into account. A run-down or substandard residence will depress the value of its neighbors. The result of these calculations is the clustering of homes by value and, since housing markets sort households by income, the subsequent grouping together of households with similar incomes. Two-million-dollar condominiums will not be built in areas with $200,000 homes but in areas of similarly priced units. Combine all of this with people's inclination to live with others like themselves, particularly (but not limited) to similarities in lifestyle and income, and to live in the best home and neighborhood that they can afford, and the result is a city with neighborhoods sorted on the basis of income and wealth: super-rich neighborhoods, middle-income neighborhoods, working-class neighborhoods, and slums for the poorest of households.[40]

Those who live in the most expensive neighborhoods will enjoy good public schools, well-maintained parks and playgrounds, sidewalks and streets in good repair, excellent trash and snow removal (if needed), and low crime. And, if they have retail areas, the shops will offer top-quality goods. When this array of high property values and services is threatened, its residents will mobilize resources and social connections to assure that the status quo is preserved. Having access to legal and political skills and strong neighborhood organizations, affluent neighborhoods protect themselves, thereby reinforcing their place in the city's residential ecology.

At the other end of the spectrum are neighborhoods populated mainly by poor or near-poor households and homes much lower in value. There, fewer people own and rents are relatively low, which is why the housing attracts poorer individuals. Their retail districts have vacant storefronts and low-quality and oftentimes overpriced goods for sale. The public schools are deficient and other services are meager. Social connections exist but they do not lead to opportunities to move out of poverty. In growing cities, low-rent neighborhoods are susceptible to gentrification in which more affluent households purchase homes cheaply and renovate them, soon to be followed by developers turning empty lots into apartment buildings. The neighborhood changes and, while many current residents might stay, few new residents like them are apt to move in.

If we broaden the urban focus to the metropolitan area, we find a corresponding, uneven landscape.[41] The ecology of neighborhoods differentiated by property value and income now becomes a municipal ecology. And, like city neighborhoods, the more affluent suburbs have better public and private services (including schools), residents with numerous influential social contacts, and appreciating property values that add to their wealth. These places—Greenwich (CT) outside New York City and Lake Forest (IL) outside Chicago—are advantaged relative to the poor municipalities within the metropolis. The latter are often older suburbs either of an early post–Second World War vintage or former industrial satellites whose industry has collapsed or left. Chester (PA) adjacent to Wilmington and Lackawanna outside the city of Buffalo are two of many such places. The weak tax bases of these municipalities provide little hope that conditions will soon change.[42]

In short, cities and their metropolitan areas do not just express the concentration of wealth in their landscapes. Spatial arrangements further exacerbate these differences. Most importantly, spatial (and social) disparities are reinforced by governmental policies that privilege the already privileged, both for tax purposes and because of social connections between those who command wealth and those who make public policy.

Another of the reasons for urban wealth concentration involves the clustering of opportunities available for living well. People migrate to cities because their current residence fails to reward their capabilities and fulfill their aspirations. In response, they take their skill at playing the cello or acting or their education in a graduate school of journalism to where these talents might be better appreciated and rewarded. The cinematic version of this is the young actress who dazzles the audience in college plays and then is drawn to New York City where she becomes a Broadway star. In the city, her ability to inhabit her characters earns her praise, adulation, and financial reward. Many others, immigrants being the most prevalent, locate in cities because of the concentration of employment and educational opportunities. There, talent, education, and connections will, they hope, be acknowledged. Immigrants with language difficulties and poor educations, however, often end up in low-wage industries where benefits are meager, hours of work unstable, and opportunities for advancement few. So positioned, it becomes more difficult for them to rise to the upper reaches of an industry or launch a lucrative career.

To take advantage of these opportunities, one must be aware of them and consider them attainable. Cities contain numerous mechanisms that sort people across occupations and industries, and between those who own businesses and those who end up working for others, whether in business or in government. School systems that offer college preparation courses or training in plumbing or computer repair can have a big impact on what people decide to do in their lives. The neighborhood in which one lives and thus with whom one comes into contact also has an effect.[43] Living among storeowners or textile workers encourages a person to see these as potential occupations. Place matters.

One's ability to move about the city and enter into downtown office buildings or nightclubs, visit hospitals, or have easy access to a manufacturing district also plays a role in shaping one's understandings of how one might live. These engagements with the world of work are mediated through family background. Those who have been sent to good schools, have met a variety of successful people, been exposed to different types of activities from cooking in a restaurant to teaching at a university, and have traveled widely are presented with more choices.

As people enter into and are sorted across different occupations and work environments, they are arranged in other categories as well. One of the most important for the concentration of wealth is whether a person ends up self-employed, operating a business that she owns and from which she can extract both wages and profits, or working for others. Owning a prosperous business, one that has a significant share of the market or provides a very high-priced good or service to affluent consumers, is a primary path to becoming wealthy. Not all businesses succeed or are able to meet these thresholds. Those that do, generate income beyond what the owners need to live. This excess can be invested to increase their wealth.[44] Becoming one of a small number of suppliers of concrete to the local construction industry, the manager of a string of fast-food franchises, senior partner at the city's premier real estate law firm, or the sole business in the city making and selling high-end picture frames are only a few examples. Owning these firms is far better than working in most occupations as a wage or salary worker. Owners make money not only from managing the business but also from the value created by the employees (only a portion of which is returned to them in wages). And, if that business can dominate the market, even greater profits are possible.

Nonetheless, employees of certain industries have the potential to earn higher incomes and amass even greater wealth than business owners. Here we find individuals working in banks, financial services firms, investment companies, and large corporations where salaries for upper-level and other key personnel are significant. A chief financial officer for a big international corporation can earn millions of dollars each year and a worker in an investment bank, paid a large, year-end bonus depending on the firm's profitability, is likely to do equally as well. Doctors, particularly those with specialties, and corporate lawyers also have this potential. Heads of public authorities, foundations, and universities are extremely well paid compared to the average worker, as are professors in research universities, experienced pilots for major airlines, and business consultants. All of these people have earnings that enable them to purchase property and invest in ways that multiply their wealth.

In addition to initial advantages and geographical effects, we also have to consider social connections.[45] These connections are an initial and continuing advantage in learning about and exploiting business opportunities, job openings, and investment possibilities. Such connections come in many forms: friendships developed in college, family, business acquaintances, and social ties to government officials, neighbors, and fellow volunteers at (and donors to) cultural and charitable organizations. Moreover, they are concentrated spatially and thus easier to access in cities, particularly when they depend on face-to-face engagement. More accessible, they are more valuable.[46] Hardly automatic in their consequences, they need to be cultivated with some people subsequently becoming more networked—and thus influential—than others. A person born into a real estate family is likely to know more about real estate development, construction, marketing, and finance, and have more acquaintances among large property owners, bankers, government officials, and architects, than someone who is not. These ties can be strengthened by making campaign contributions to people running for elective office, donating to and being involved in charities, and involving others in one's projects (hoping that they will reciprocate later). Without these connections, a person is isolated and, so isolated, is less likely to encounter wealth-enhancing opportunities.

The concentration of wealth is further exacerbated when those with high incomes have more money than they need to live comfortably. The surplus is then invested in businesses (buying stocks, for example), multiple properties (a home in the city and the countryside or part-ownership of a small office building), expensive automobiles and paintings, and private schools for their children. For those who have good connections and can pay for sound financial advice, these investments are likely to yield substantial rewards and expand the family's wealth. Once again, initial advantages are translated into further advantages. And while the rich might invest poorly and lose assets, catastrophic losses are unlikely given their access to professional advice. The myth that anyone can become rich in the United States is overshadowed by the reality that once a family has become wealthy, it is likely to stay wealthy. Those lacking wealth cannot take advantage of investment opportunities or develop the requisite educational skills and social connections. Consequently, they fall further and further behind those who can. Rather than a "land of opportunity—where hard workers from any background can prosper," the reality in the United States "is far less encouraging" with social mobility about average for democratic countries with market economies.[47] The city sets out the possibilities and those with initial advantages and connections use them to access resources, strengthen their capabilities, and concentrate wealth.

The wealth to which these initial and subsequent advantages lead is protected and stabilized by governmental policies. Laws regulating private property, business taxes, and corporate status; the relationship between management and labor; restrictions on labor organizing; zoning regulations that bolster property values; inheritance laws; and tax policies constitute a framework that perpetuates the concentration of wealth. Wealth does not have to be shared. Widely accepted in the United States is that those who succeed at becoming rich can do whatever they wish with their fortunes, subject to certain taxes and minimal regulations. At times, the rich might feel embattled, but the evidence suggests that their affluence is hardly at risk of being lost. A dominant ideology—freedom, unlimited opportunities, and rewards for hard work—justifies the concentration of wealth and reinforces the self-serving notion that the wealthy have—by themselves—earned their riches.

A number of government policies counteract these tendencies. The federal income tax system is more or less progressive, drawing a larger percentage of a richer person's income. Combined with greater assets, this means that the rich pay more in taxes, as they should. Inheritance laws dampen the intergenerational transfer of wealth. As regards inequality, government programs exist to improve the lives of those who are struggling to live well. School lunch programs for poor families, tax credits for low income households, subsidized housing and health care, retirement assistance, compensatory education, unemployment insurance, and special services (for example, assisted transit for the elderly) are just a few of these efforts. They are funded and managed by various levels of government with city governments frequently involved.[48] Together, though, they hardly lessen the concentration of wealth or significantly decrease the gap between the rich and the poor. Even as the government addresses the consequences of wealth concentration, it simultaneously subsidizes corporations and exacerbates that concentration. Farm subsidies go disproportionately to big agricultural producers, tariffs are imposed on imports to protect domestic industries, and large banks and corporations are kept from failing during severe recessions. Weak federal inheritance laws allow large fortunes to be passed from generation to generation.

Many of these governmental efforts to protect the wealthy and their wealth-generating assets emanate from the federal government; others originate in state governments. They are appropriately viewed as non-urban in origin, though all affect cities and those who live there. Nevertheless, city governments are not absent from this story. Their tax policies, though, are hardly redistributive. Income taxes are more progressive than real estate or sales taxes, but relatively few cities have them. In places like New York City, Birmingham, Philadelphia, and St. Louis that do, the income tax rates are relatively low compared to federal rates. In 2014, the rate was 1.25% in Wilmington (DEL), 1.85% in Bowling Green (KY), and 1% in St. Louis.[49] These flat taxes are regressive in their consequences: the proportion of one's income paid to the government does not rise with either income or wealth. Sales taxes are even more regressive in that less affluent households spend a higher proportion of their income on goods and services than more affluent households.

In addition, zoning and land use regulations protect property from being devalued by the intrusion of inappropriate users as when a nightclub is blocked from opening in a high-end retail district. The city's best schools are often in its most affluent neighborhoods. Policing is disproportionately targeted on minority youth, adding another barrier to upward mobility. Sidewalks and parks are likely to be less maintained in poor rather than rich neighborhoods, thereby discouraging middle-income households and developers. On the other side of the ledger, city governments create value for property owners and investors when they increase the allowable densities for development or open up opportunities for building apartments or hotels in industrial areas. In these many ways, cities concentrate wealth in the hands of a minority of the city's residents.

Cities not only enable the generation and concentration of wealth, they also do the same for poverty. Wealth and poverty, though, are not two separate and independent states of affairs. Rather, the concentration of wealth engenders inequality with high concentrations of poor people often living in the same cities as concentrations of affluent households. Poverty is a correlate, even if an imperfect one, of affluence.

Concentration of Poverty

Poor people have lived in the country's cities since European colonization. And, despite numerous efforts by charitable organizations, churches, and governments to eliminate poverty, poverty remains a reality in contemporary urban areas. Poverty can also be found in rural areas and is just as devastating to people's lives. It differs, however, from urban poverty in a number of ways, the most important of which is its spatial juxtaposition and functional relationship to wealth. One commentator has noted, beginning somewhat breathlessly: "Cities are extraordinary economic engines of wealth and innovation, but this same mechanism can cause terrible inequality and poverty."[50]

In 2012, an estimated 15 percent of the US population—47.1 million people—were living in conditions of poverty. Notwithstanding the spread of poor people into the suburbs, poverty had become (over the decades) more clustered and concentrated in distressed and high-poverty urban neighborhoods: "the poor are over-represented in the central cities of every one of America's metropolitan areas."[51] In these areas, 23% of the poor lived in high-poverty neighborhoods compared to 6.3% of the suburban poor. Moreover, a higher percentage (18.2%) of residents of cities were poor compared to residents of the suburbs (9.5%).[52] Poverty, moreover, is present not only in cities whose economies have faltered. Wealthy cities have significant percentages of poor people as well. In 2013, Boston was simultaneously the sixth wealthiest large city and the tenth poorest. Even in growing cities, poor people are still an embarrassing portion of the population. For example, El Paso, which had a 15% population increase between 2000 and 2010 and is still growing, was nonetheless the eighth poorest large city in the United States in 2013 with nearly 31% of its households earning less than $25,000 per year.[53] (See table 2.5.)

Admittedly, poverty is not a condition peculiar to cities; it is a national (and even global) phenomenon. It originates in an economy that fails to generate a sufficient number of well-paying jobs and that allocates a large portion of the population to low wages or intermittent work. Many people are unable to become sufficiently wealthy to weather difficult times or to live well during retirement. In numerous low-wage industries, labor is unorganized and fails to bring enough pressure on owners and managers to assure high wages and adequate benefits and protections. Those with initial disadvantages thus find

Table 2.5 **Poorest and Wealthiest Cities in the United States, 2014**

Poorest Cities	Percent Residents Earning Less Than $25,000/Year
Detroit	48.0
Milwaukee	36.5
Philadelphia	36.4
Memphis	34.9
Tucson	34.8
Wealthiest Cities	**Percent Residents Earning More than $150,000/Year**
San Francisco	23.4
San Jose	22.6
Washington, DC	19.0
Seattle	16.2
San Diego	14.8

Source: Ryan Childs, "These Are the Poorest Cities in America, *Time*, November 14, 2014, and Ryan Childs, "These Are the Wealthiest Cities in America," *Time*, October 30, 2014.

themselves struggling to live decently or to gain access to the kind of education or develop the skills that will move them to higher-paying positions. Moreover, they are pushed into housing markets where a stable home life is near-impossible.[54] Immigrants with little education and having to learn a new language and culture are particularly disadvantaged. The United States does not have powerful mechanisms that move people into high-wage jobs, maintain a living wage across industries, dampen wage inequality within industries and corporations, provide enough well-paying jobs to keep all working-age adults and their families out of poverty, or assure adequate housing.

African-Americans, other people of color (for example, Latinos), and Asian, Middle Eastern, and African immigrants face discrimination as well as exploitation. Despite anti-discrimination legislation, these individuals have difficulty being hired, gaining access to neighborhoods with good schools for their children, and positioning themselves in social networks that lead to opportunities for advancement. Minority individuals are much more likely to be poor than those in the white majority and much more likely to be concentrated in urban areas. In 2008, African-Americans and Latinos constituted 67% of the poor in major cities but only 24% in new suburbs and 19% in the exurbs.[55]

The disadvantage of being a minority is exacerbated for the elderly. While African-Americans were about 9 percent of the population in 2008, they represented 21 percent of the elderly poor. And even though poverty rates for the elderly are higher in rural than urban areas, the elderly still constitute a significant portion of the urban poor. In Beverly Hills, one of the most affluent cities in the country, 4 out of 10 poor people are senior citizens. On the other coast, in New York City, the elderly are a growing segment of the population with one in five citizens over 60 years of age considered to be poor. The poverty rates of the immigrant elderly are even higher and elderly women are much more likely to be poor than elderly men. For cities where in-migration is meager and the population is shrinking, thereby increasing the proportion of elderly residents, this is a serious problem.[56]

The homeless—a small but important category of poor people—are heavily concentrated in cities. One estimate has about 85 percent of all homeless living in cities with one out of five in either New York City or Los Angeles—Los Angeles being labeled the "homeless capital" of America. In 2013, New York City had the largest number of homeless, but Los Angeles had the highest number of homeless living outside of shelters; that is, living on the streets. Seattle is another city with a large contingent of homeless. Note that all three are relatively affluent cities.[57]

Numerous factors, both urban and non-urban, make life even more difficult for the poor. High on the list in terms of effect are insufficient governmental supports to overcome initial disadvantages and compensate for low-wage or intermittent employment along with geographical conditions that diminish access to resources and connections and thereby stifle capabilities. The federal government does provide subsidized housing and health care (though not to all) and financial benefits to retired workers. This has reduced poverty, most dramatically among the aged.[58] At the local level, however, governmental policies are too anemic to shrink income disparities. Children born into low-income families are inadequately served by school systems. Laws supporting union organizing are weak. Housing subsidies do not extend to those most in need. Food assistance is far from sufficient. Despite numerous programs at the federal, state, and even local levels of government, poverty persists and is a condition from which an embarrassing, large portion of the country's residents suffer.

Where people live also contributes to poverty. Economic activity is unevenly distributed across the country, within metropolitan areas, and in cities. Rural areas where mines have closed, industrial suburbs and cities where the dominant manufacturing firm has left, and neighborhoods where the only nearby jobs are in small, retail establishments, street vending, or informal activities offer few opportunities for living well. Such places fail to attract new investment sufficient to generate the jobs needed to reinvigorate the local economy. For those who have lost their jobs and have few savings, living there is unlikely to improve their quality of life. Moreover, they often do not want to relocate and leave family and friends behind; consider themselves too old to move elsewhere or take up another occupation; or simply lack the resources to do any of these things. The result is an increase in the number of poor people in these places and the further concentration of poverty.

Relocating, though, makes sense, particularly if the poor live in a shrinking city from which most of the middle-class and likely all of the wealthy have departed. Poor people are better off in cities with a wealthy and highly educated population. There, the schools are better, health care is more accessible, and local governmental expenditures on public services are higher. Opportunities are more numerous. And, to the extent that life expectancy is diminished by poverty and poor health, living in more advantaged places is likely to increase it as well.

Life expectancy is highest in New York, Santa Barbara, and San Jose and lowest in Gary (IN) and Tulsa.[59]

Where a household lives within a city also has implications for poverty. Consider San Antonio, "the most spatially unequal city in the country."[60] Comparing its poorest areas with its wealthiest, the differences are striking. In the former, 4 out of 10 households lived below the poverty line in 2013; in the latter, one of 25 did. The poorest areas have few people with a high school diploma, older housing, more unemployment, and a declining economy with businesses leaving rather than entering the neighborhood. Most telling, growth in the city's most affluent areas is "doing little to raise the fortunes" of the most distressed areas.

Involuntary residential segregation by income and race plays a central role. The only housing options poor people have are in neighborhoods with other poor people. The odds of finding an affordable apartment in an affluent neighborhood are against them. And for minorities, if their income does not force them to live in a marginal neighborhood, it will when combined with discrimination in the housing market. Comparing white and African-American families with similar incomes, an African-American family is more likely to live in a poor neighborhood lacking day care options, good schools, and playgrounds. This racial gap (regardless of income, remember) varies across cities, with the largest gaps in Milwaukee, Newark, and Gary and the smallest in El Paso, Riverside (CA), and Albuquerque.[61] In these segregated neighborhoods, African-American families have limited access to jobs, are deprived of decent and affordable housing as well as opportunities to amass wealth from homeownership, and constricted in their social connections. Their children are channeled into weak schools that lack the capacity to adequately educate them. That they might choose to live with others like themselves and close to their families and find social support there does not obviate the fact that being confined with other poor people exacerbates their initial disadvantages.[62]

Social connections are particularly important. A person living in a poor neighborhood with no ties to anyone who is wealthy is disadvantaged in the quest for riches. Her acquaintances have knowledge only about low-wage jobs, few business opportunities are forthcoming, and advice on how best to invest or save for unexpected events is scarce. For people on the city's geographical and social margins, what social connections they have—and they have many that support them and enrich their lives—are less apt to magnify initial advantages or generate wealth. Even pooling resources and collaboration is difficult given the few resources they possess and the lack of investment opportunities where they live. Their social connections—mediated by proximity and density—concentrate their disadvantages socially and geographically.

When people with few marketable skills, weak educational backgrounds, and minimal resources do move, they do so to both growing and declining cities. Growing cities are attractive because business and job opportunities are expanding and growth promises that they will be able to improve their current prospects. The population is increasing and businesses are hiring. With many people doing well financially, numerous service jobs—nannies, delivery boys, waitresses, store clerks, non-union construction laborer—exist for those with little education and weak language skills. Once established in an entry-level job and, most importantly, in a decent neighborhood, the odds of moving to better and better positions improve.

People are also attracted to cities that are losing population and businesses and where property values are falling. Camden, New Jersey, one of the poorest and most crime-ridden cities in the country, has lost population but not emptied out. Many poor households move into the city for the cheap housing and access to suburban jobs. People leave if they can, while others settle there because it is all that they can afford.

Further concentrating poverty in the city is the clustering of businesses and jobs in terms of wages. High value and high-wage businesses are almost always located in the central business district of the city or in smaller satellite ("edge") cities in the suburbs. And while the central business district is often accessible from poor neighborhoods, residents there are unlikely to have the social connections to obtain the better-paying positions. Because many good-paying jobs are in the suburbs and edge cities, and because public transit is usually absent, commuting almost always requires an automobile. This is a major barrier for disadvantaged households. Overall, few income-producing business opportunities and well-paid jobs are available to the working poor where they live. This blocks their advancement to middle-class status and confines them to these areas. Even if cities are not the main culprits in generating poverty in the United States, they are significant contributors to its concentration and perpetuation.

Wealth and Poverty

The dynamics that generate and concentrate wealth have, as one of their effects, the generation and concentration of poverty. The reverse (generation and concentration of poverty leads to generation and concentration of wealth), however, is not the case. Wealth and poverty are asymmetrically related. Consider these ways in which the concentration of wealth exacerbates poverty.[63]

In the world where wealthy people own and manage businesses that prosper and in cities where many such entrepreneurs and investors live, two consequences ensue. The first is that business opportunities are quickly taken up if they show signs of significant growth and robust profitability. Or, to put it differently, if there is money to be made, the already-wealthy will be involved. This is a very common experience for start-up firms needing capital or growing rapidly. They attract major investors who, in return for their infusion of capital, take partial ownership and control. Consider another example. A young, immigrant entrepreneur develops a highly acclaimed clothing store in a marginal neighborhood and soon thereafter other entrepreneurs open similar shops nearby to capture some of the cachet of this recently discovered "destination." In short, the wealthy can easily identify and exploit business opportunities. By doing so, they dominate many of the opportunities for prosperity and make it difficult for those with little wealth to become wealthy.

The second consequence concerns the ways in which wealth generates low-wage jobs. Large and prosperous businesses and institutions derive a portion of their profits by minimizing their wage bill. The secretarial and cleaning staffs of law firms, the maintenance workers in hospitals and universities, the clerks in insurance companies, and the restaurant workers, ushers, restroom staff, and parking attendants at professional sporting events are all paid much, much less than the executives of these firms. The profits of these entities are as high as they are in part because labor costs are held down. For some of these workers, their wages might be sufficient to support a family, if their partner also has a job. For others, particularly those with only part-time and seasonal work, it is not; they hover close to the poverty line. Wealthy businesses do not generate poverty directly. Rather, they maintain inequalities of income that, in turn, lead to other inequalities.

The demand that the wealthy have for service workers has similar consequences. Wealthy people have money to hire people to do tasks that non-wealthy people do for themselves. A family in a high income bracket often has a cleaning person, a nanny for the children, and a dog-walker. The super-rich have their own cooks and chauffeurs, along with gardeners and caretakers for their country estates. The wealthy frequently dine out at restaurants where the back-of-the-house staff receives minimum wage (if that), take taxis whose drivers struggle daily to make a decent living, and attend gyms where the staff earns just enough to satisfy basic needs. Then there are the carpenters who rebuild the kitchens and the doormen who watch over the building in which the wealthy live. Affluent households desire and pay for a range of services and, while this is not a significant factor in expanding and concentrating the poor, it maintains their wealth and status and, by doing so, stabilizes the gap between them and the less fortunate. Without doubt, these purchases also create and support jobs. Most of these jobs, however, are low-wage positions.

More directly related to the coexistence of wealth and poverty in cities are the dynamics related to place itself. One of them involves the efforts that the wealthy make to protect their neighborhoods from encroachment by unwanted land uses or people (for example, a halfway house for abused women, a car impoundment lot) and thus from any change that might threaten the value of their property or the types of people who live there. One consequence of this is to shift these activities and people to neighborhoods where residents are less politically connected and have fewer resources to resist. This solidifies the hierarchy of neighborhoods within the city while often burdening the residents of poor neighborhoods with unwanted facilities that generate noise, air pollution, and traffic.

The second way in which place dynamics affect inequality is more convoluted but nonetheless real. The nature of capitalist real estate dynamics is to be constantly restless, searching for new investment opportunities and moving on when buildings become old and obsolete and profits falter. The relentless outward and upward growth of the city is, in part, a consequence of this tendency, with developers opting to invest on greenfield sites rather than rehabilitate where they have already built or replace a profitable office building with one taller and even more profitable. One result is a landscape with high and low property values: affluent and poor neighborhoods, affluent and marginal retail districts, and affluent and low-end office areas. This discrepancy in values is beneficial for real estate investors, particularly when cities are growing. Money can be made by investing in an area that has been devalued but has the potential to be revalued.[64]

Lastly, and to repeat a point made earlier, cities with wealthy populations and growing economies attract poor people. Poor people migrate to cities because, by doing so, they improve their prospects for financial and social advancement. The dynamics of wealth, politics, and the ecology of neighborhoods nevertheless pose formidable barriers to upward mobility, despite the many opportunities that are made available. Cities enable both great wealth and enduring poverty. They concentrate the wealthy and the poor and they stabilize inequalities. Cities "are engines of prosperity and inequality in equal measure, and when the inequality tips poor they look unsaveable; when it tips rich, they look unjust."[65]

THREE

Destructive, Sustainable

Every day, people in Chicago consume immense quantities of food that arrive from nearby farms, large agribusinesses further afield, and producers around the world. Oranges are sent from citrus groves in Florida, chicken parts from processing plants in South Carolina, bananas from growers in the Dominican Republic, olive oil from Italy, wine from the Loire Valley in France, and grapes from Chile. Cab drivers, automobile commuters, and truck drivers purchase gasoline that began as crude oil in Canada, the Gulf of Mexico, Venezuela, and Saudi Arabia and was refined in Louisiana and Texas. Furniture is shipped from Sweden, machinery from Germany, train cars from Canada, and clothing from Turkey. For the people of Chicago to live well and its visitors to eat and sleep, conduct business, and enjoy the tourist experience—for the city to function—raw materials and finished products have to be imported from places beyond the city limits.

Chicago is unexceptional in this way; all cities draw resources and commodities from far-flung places. Their ecological footprints—the land area devoted to their care and feeding—exceeds by many magnitudes the area occupied by their residents. For very large cities like Los Angeles and Dallas, their footprints are vast. The disposal of the human waste, plastic food packaging, construction debris, and discarded paper and the water runoff and air pollution attendant to their functioning further expand the places brought into the environmental frame. Outmoded computers are sent to China for recycling, air pollution drifts across the region, and used clothing is shipped overseas to developing countries. The ecological impact of cities reminds us, once again, that while cities might be defined by their political territories, they are hardly contained within them.

In short, cities use more of the environment than they occupy. They overflow their boundaries. When they grow, their reach extends even further and, as new cities arise, even more land and resources are drawn on for sustenance. Jungles in Brazil are turned into pastures, the tops of hills in West Virginia are removed so that coal can be strip-mined, and shoe factories built in Vietnam draw more and more people into the cities from the countryside. The extraction of resources, the use of land for grazing, forestry, and crop production, and the manufacturing of goods are all intensified. When it comes to the natural environment, cities are greedy. They "consume 75% of the world's energy and produce 80% of its greenhouse gas emissions."[1] As the population of the world continues to grow (exceeding 7.4 billion in 2016 according to the United Nations) and urbanization increases (exceeding 54 percent of the world's population), it is difficult to accept calmly that more and more and larger and larger cities actually contribute to a sustainable world. Can one really argue that large cities protect ecological habitats, preserve valued landscapes, oceans, and glaciers, and tread lightly on nonrenewable resources such that the earth can support future generations?

Yet, cities are touted as saviors of the environment, our only hope for sustainable urbanization. As one observer has commented, "Cities represent the best chance of realizing the aspiration of global sustainability."[2] For many observers, they are the only form of human settlement with the potential to minimize energy use, encroach as little as possible on the natural environment, and integrate human activity with ecological settings. Purportedly, they are also the only settlement form able to absorb an ever-expanding population. Low-density suburbs, small towns, and rural villages fall well short of these goals when compared to dense, compact cities. The call from popular commentators and scholars alike is for cities to be sustainable and resilient. The first protects, as much as possible, the natural environment for future generations, while the second keeps them from falling apart in the face of disaster and disruption. Together, sustainability and resilience establish the basis for additional urban growth. Having these qualities, large cities are neither a threat to the environment and, by extension, to human functioning, nor a source of insecurity. Rather, they are the places, the sole places, where sustainability makes sense.[3]

Which is it then? Are cities a curse on land, air, water, and mineral resources, not to mention on animals, birds, plants, and creatures of the sea—an insult to nature? Contrarily, are they the best alternative that humans have for protecting the natural environment and its many resources and ecologies? What can be done in the face of continued population growth and the unrelenting urbanization that together fuel consumption and deplete and degrade the material world? My argument, as expected, is that they are both and can never be either one or the other. Cities are always destructive; they cannot be otherwise. At the same time, cities hold out the promise of sustainability.

Given the subsistence level at which a significant portion of the world's population lives—896 million people below the international poverty line in 2012—coupled with the rapacious appetite of those who reside in high-income countries, more and more of the material world will have to be consumed—not less—in order to reduce those living in poverty to manageable numbers.[4] The only reasonable path to sustainability, though, seems to be to decrease the ecological footprint of humans and limit how much each of us, and each of our technologies from factories to gas-powered vehicles to electrical infrastructure, takes from the earth. Such an objective can be seriously considered only if we account for the need to improve the quality of life among the multitudes who live a bare and mean existence. For the cynical among us, even a modest stabilization of human presence seems unlikely. As cities grow, they require more of the world for themselves, regardless of how sustainable this might be. In fact, the more sustainable cities become, the stronger the justification for allowing increased urbanization with more, not less, ecological destruction the result.

This chapter is organized around the two themes signaled by its title: the negative impact that cities have had and continue to have on the environment and the prospects and quest for sustainability. To understand the tension between the two, we need to acknowledge that the country's cities are solidly entrenched—8 of 10 people living in urbanized areas—and the benefits of urban living, at least for now, are widely publicized and embraced. No one that I know of, moreover, has imagined an alternative and equivalent form of human settlement. Consequently, we can neither return to a state of nature, and innocence, nor abandon the cities. Our only hope is to become better at mediating between the city's environmental destructiveness and its potential for sustainability.

Cities and Nature

The beginning point for any negotiation of this contradiction begins with recognition of the inseparability of culture and nature and, thus, the indivisibility of cities and the natural environment. The modernist conceit is that humans are a species qualitatively different from and superior in many ways to other living things. Humans have free will, interact with each other using elaborate languages, collaborate to produce steam engines and cities, are able to imagine the consequences of their intentions, and often act emotionally (even though they are ultimately in control of their baser urges). Plants, animals, birds, and insects ostensibly lack similar qualities and these differences enable humans to dominate them. This argument justifies thinking of the world wholly from a human perspective. The belief, however, is false. Culture and nature are not separate and distinct realms. Anyone who interacts with animals, for example, soon realizes that they have intentions and are capable of learning, much like humans. We are not so different. The modernist conceit is based on a false distinction and, partly because of this, cities should not be considered a displacement of nature but rather a collaboration with it.[5]

For the most part, cities are set within the landscape and tightly joined to the natural world on which they depend. Without the air and water that ecologies provide, humans would not be able to survive. Rock and soils support their buildings and roads and provide the materials for making walls, floors, and pavements. Plants, animals, and insects are abundant and, in almost all instances, an unavoidable presence. Set within "nature," cities have not, however, left it intact. Trees are removed and hills flattened to make sites for homes and office parks. Rivers are dredged and their banks shaped so that docks can be built. The ecologies of predators (such as wolves) are disrupted and even destroyed. For centuries, and as humans built their settlements, wetlands that had once provided food and shelter for birds, fish, and other aquatic creatures were filled with gravel or, worse, with the ashes from incinerators, construction debris, and the discarded objects of daily life. Factories and power plants pollute the nearby streams and soils on which they are built. Plant colonies are uprooted, animal and bird migratory pathways disturbed, and rivers and streams dammed so as to create reservoirs and navigable bodies of water. As a consequence, fewer wild animals are present, even though deer and black bears in search of food now frequently wander into the suburbs of Denver, New York, and other cities.

As cities evolve, new plants, animals, and land forms appear.[6] Wild berries and sage brush are pushed aside by lawns, ornamental plants, and shade trees. Domesticated pets—dogs, cats, goldfish, parakeets—replace the foxes, snakes, and rabbits that had previously inhabited these places. Rats and pigeons multiply as they adapt to city life; hawks build nests on apartment buildings to survey the sidewalks and parks for prey and to protect their young. Certain insects—mosquitoes, spiders, and cockroaches—become more prevalent. Ants, houseflies, and bedbugs invade residences and businesses. Bird species become fewer as their habitats are displaced by homes and factories. Novel land forms (such as outdoor amphitheaters and ponds) are placed in parks with many of these bodies of water and open fields becoming stopover destinations for migratory birds. New surfaces appear: cobblestone streets, asphalt roads, crushed stone walkways, and permeable concrete alleyways. Nonnative plants (lilac bushes) and trees (ailanthus) are introduced. Much of nature is displaced by the city, but nature is still present.

To establish these accommodations and collaborations, numerous technologies are developed and deployed.[7] They are built so that humans can communicate with each other (as with cell-phone towers) or travel from place to place or engage with (while also exploiting) nature and its many ecologies. At times, these technologies are benign; at other times, destructive. Funiculars are built so that humans can traverse steep slopes, while (visually less intrusive but nonetheless problematic) pesticides are used to control the mosquito population in the wetlands along the lakefront. Dams, aqueducts, and sewer systems are used to store and channel water. Treatment plants make water potable. Devices are attached to the smokestacks of factories and utility plants to clean pollutants from the fumes that are being discharged; catalytic converters on automobiles and trucks perform a similar function. Waste disposal systems combat noxious smells by removing decaying flesh, food, and feces from streets and backyards. Traps are set for mice and rats. Exterminators spray chemicals to eradicate cockroaches and other insects.

Areas of parks and playgrounds are designated as dog runs. Regulations are posted on their entry gates and the urine-saturated soils are periodically replaced. Laws are passed to control which animals can be kept as pets and how they can be disposed of when they die. Walls and roofs protect humans from the rain, the sun, and the wind. Special foundation systems have been invented to enable tall buildings to rise skyward and serve as office space for corporate executives. Air quality monitors indicate when automobile traffic should be limited. Temperature signs remind us of what to wear and weather forecasts warn of imminent flooding or sandstorms. In the city, humans engage nature directly and through the intermediaries of technologies.

This interaction of technology, nature, and humans is central to the founding, growth, and development of cities. San Francisco is a typical example: nature and the city (supposed opposites) are intertwined.[8] The city could not have grown without nearby farms to produce foods for its residents. As the city grew, however, it did not simply push these farms further away, but absorbed them into its periphery. Watersheds were designated and pumping technologies used to bring water from afar, thus integrating them into the urban fabric. Construction required building materials and, as a result, brick-making operations, concrete plants, and quarries (to mine sand and gravel) came to be interspersed throughout the region. The bay, a defining element of the San Francisco landscape, was dredged for minerals, used for commercial fishing and the harvesting of shellfish, and filled with rocks and soils to make land for new development. Pollution from pesticides and manufacturing along with the runoff from roadways further transformed the bay. None of this was inseparable from local and even supra-local politics as different groups vied for place and profits and worked to make San Francisco a city attuned to their needs. Despite the unavoidable dependence of humans on nature, as geographer Richard Walker has noted, "no city will ever be reconciled with the countryside."[9] For him, this is inevitable when growth is a dominant goal and capitalism the dominant form of political-economy.

Technologies are not just ways of adapting nature to human purposes.[10] Humans also have to adapt to and collaborate with nature; they have to share the city with it—coexist. If cities like Las Vegas, Los Angeles, and Phoenix that lack nearby sources of water are to continue to grow, they must find ways to accommodate to water scarcity using technology to bring water to them and finding means to utilize and recycle efficiently the water that they have. The residents of Denver have to accept that winter will bring major snow storms that will slow travel, despite fleets of snow plows. The suburban residents of many metropolitan areas can scarcely ignore the non-domesticated animals (such as deer, coyotes, black bears, and turkeys) that wander into their midst and disrupt their lives. Office towers have to be designed to minimize the disorienting effects that their glass facades have on migratory birds, thereby preventing bird deaths and damage to the buildings. And, as most residents of cities are well aware, stepping around that flock of pigeons on the sidewalk might well be preferable to walking through them.

To think of the city as solely the realm of humans is to inhabit a fantasy world. Such cities simply do not exist. Cities are not "for people" alone. Acknowledging this is the first step in recognizing that the city is an accommodation and collaboration with nature. And although such arrangements vary greatly in the benefits they confer on each of the parties and the consequences they have for environmental sustainability, they are no less important or consequential because of it. The extra benefit of such a perspective is to undermine the hubris of humans that portrays them as masters of the universe.

Dominating Nature

We can grasp the actual as well as the potential destructiveness of cities by reflecting on them in a historical context. Their destructive impact was particularly striking before the passage of environmental laws in the mid-twentieth century meant to preserve ecological sites, mandate levels of air and water quality, protect endangered species, and encourage the use of renewable energy sources. Three cities, at different points in history, will serve as examples: Jamestown, Pittsburgh, and Phoenix. With this understanding in place, we can then turn to two of the ways that policymakers and researchers think about the intrusion of cities on nature: urban metabolism and the ecological footprint. Tying them together is the undeniable presence of urban sprawl.

The destructiveness and sustainability of human settlements has a good deal to do with scale; that is, with how large and dense these settlements are and the numbers of people who work, live, and play there. Particularly important are the technologies that support these activities. They enable humans to grow crops, make use of natural resources (such as water sources), produce commodities from buildings to clothing, and dispose of waste. Many of the technologies used in the countryside for raising crops and mining stone, moreover, were created in cities. This is one of the reasons that the famous urban commentator Jane Jacobs argued that "rural economies, including agricultural work, are directly built on city economies and city work."[11] By this scalar logic, when settlements are sparsely populated, their inhabitants can live by using renewable resources (for example, harvesting trees for fuel), take fish from streams without depleting the stocks, or farm the land so as to maintain the quality of the soil. The environment is not irreparably harmed. As one historian commented about one specific environmental problem: "Outdoor air pollution of any serious consequence came only with cities."[12]

Contrary to this logic, small and low-density human settlements are not necessarily a path to environmental sustainability. The ancient city of Ur, for example, "was responsible for the deforestation of a large area of Mesopotamia and experienced shortages of water due to intense agricultural cultivation."[13] An equally complicated set of relationships has characterized subsequent settlements. Take, as an example, Jamestown in Virginia in the years just after its founding in 1607.

Led by Captain John Smith of The Virginia Company, an entity chartered by King James as a profit-making enterprise, over 100 English settlers landed on Paspahegh island in the Powhatan (later, James) River and proceeded to build a fort along with small, thatched houses in which to live.[14] The expectation was that the settlers would trade with the natives, grow their own crops, fish and hunt, and send commodities such as fish and salt back to England. In contemporary terms, the settler community would be (by necessity) sustainable. The Algonkion-speaking tribes (known as the Powhatan) were not particularly welcoming however. They had clashed with earlier English expeditions and were suspicious of invaders. Nevertheless, they did share two of their main crops—maize (corn) and tobacco—with the newcomers. The strained relationship led to conflict with periodic killings on both sides and a massacre by the Powhatan in 1622 that left 347 settlers dead.

Past encounters between native Americans and settlers were not the only reason for the conflict. The settlers also failed to grow sufficient quantities of food and their cultivation of tobacco for trade with England used valuable agricultural land and diverted their energies from food production. Consequently, and despite the abundance of game, fish, fruit, and berries in the area, the early settlers were constantly on the verge of starvation. Seemingly incapable of planting and harvesting food to meet their needs, they were also burdened by poor water supplies and the diseases attributed to nearby swamps and marshlands. Between 1606 and 1624, 7,289 immigrants landed in Jamestown and 6,040 died of malnutrition, disease, or violence. "In the winter of 1609–10, five hundred colonists were reduced to sixty."[15] In response, the settlers took to raiding the native villages for food. What partly discouraged the settlers from growing food crops and where settlers were successful was with the lucrative trade in tobacco. This crop, though, led the Virginia Company to expropriate more and more land for tobacco plantations, thus further encroaching on lands the natives used to farm, hunt, and fish. This expansion and the constant attacks on the natives "irrationally destroyed the supplies of grain and fish sorely needed by both races."[16]

Until the English settlement, the Powhatan had lived in small villages, engaged in sedentary agriculture, and hunted and fished. Their presence was sustainable; there was an ecological balance. The Jamestown settlement changed this. The growth of tobacco as the dominant (settler) crop depleted the soil of its nutrients and led to further encroachment on the fields and forests. The growth of the settler population as a result of additional immigration drew down the stocks of fish, birds, and animals, while the need for wood for fuel and building materials depleted the forests. John Smith in 1629 wrote that "James Towne [has] most of the woods destroyed, little corn there planted, but all converted into pastures and gardens."[17] In addition, the ecological footprint of the area was significant in that the settlers mostly relied on England for tools, utensils, and items to trade with the natives. And, it expanded as the number of residents grew. The ecological footprint was further extended into the interior as tobacco plantations were established and settlers migrated westward. The story of one of the country's earliest European settlements, and thus of its earliest European towns and cities, is hardly one of environmental sustainability or sympathetic coexistence with the region's social ecology.

By the early twenty-first century, few such seemingly self-sufficient and small settlements exist in comparison to the vast number of cities whose ecological footprints extend well beyond their political boundaries. Officially designated cities in the United States in 2000 ranged in size from Carson City (NV) with 52,460 residents to New York City with 8,008,280 residents. Their land areas varied from Hoboken's (NJ) one square mile to Anchorage's (AK) 1,697 square miles. And, their densities started at Anchorage's 153 people per square mile and ended at Union City's (NJ) 52,970 people per square mile.[18] When cities becoming bigger and metropolitan areas expand in land area, the human population makes greater and greater use of the land and raises the potential for environmental damage.

Cities are not small villages of low density relying mainly on technologies that use only renewable resources.[19] Large and dense, with almost none of their residents engaged in farming, addicted to fossil-fuel energy sources, and needing to import large amounts of food and materials, they cannot live lightly on the land as could Native Americans. The industrial cities of the early twentieth century are prime examples of the way in which urbanization and technologies severely compromised the natural environment. Changes in production techniques, financing, and management practices enabled business owners to expand their firms and produce for mass markets. Small workshops were turned into large factories and, as markets grew geographically, it became more and more important to locate these factories adjacent to ports and near rail hubs. People flocked to the city, both to find employment and because the mechanization of agriculture was making small-scale farming and self-sufficient living less and less viable. Business owners and households required services such as paved streets, fire protection, and public markets. The cities grew, simultaneously becoming denser and expanding across the land.

This urbanization occurred in the absence of regulations that now exist to protect the natural environment. Air and water pollution were rampant. Soils were contaminated with the by-products of manufacturing. Wetlands were filled so that warehouses could be built next to docks. Animals were chased away; forests were denuded. Housing was built adjacent to factories and both were built in the absence of building codes and safety regulations. Wells were not inspected. The result was cities susceptible to the spread of disease and prone to destructive fires that spread from one flammable property to another. In the early twentieth century, it was still acceptable to keep pigs and goats in the city and the use of horses to draw carriages and carts contributed significantly to the unsanitary condition of the streets. The industrial city brought extensive environmental destruction.

Consider the case of Pittsburgh. Even though this is an historical case—and an example of behaviors and conditions that would not be tolerated today—it illustrates how cities engaged with the natural environment during a period of large-scale industrialization. Pittsburgh exemplifies the environmental conditions that plagued cities that had large, heavy-manufacturing sectors (such as automobile and steel production) during the late nineteenth to early twentieth centuries.

From the mid-1800s onward, Pittsburgh's prosperity came from the abundance of coal in the region and from rivers that allowed iron ore to be brought down from the north and steel to be sent by barge to markets to the east and south. Mining and iron manufacturing became critical industries. As the 1900s unfolded, steel manufacturing rose to a dominant position and massive, integrated plants were built along the riverbanks. Steel was joined by other large manufacturers: railroad equipment, metals fabrication, and electrical equipment among others. Air, water, and soil pollution were three of the consequences. The city's hilly topography pushed these factories onto the rivers' floodplains and into the river valleys with railroad tracks, rail yards, and shipping facilities located there as well. By the early twentieth century, and in the absence of environmental regulations and effective land-use controls, the city had become environmentally degraded. As early as 1866, one commentator wrote that "quiet valleys have been inundated with slag, defaced with refuse, marred by hideous buildings. Streams have been polluted with sewage and waste from the mills. Life for the majority of the population has been rendered unspeakably pinched and dingy."[20]

Pittsburgh's many factories, its reliance on coal for heating and as an energy source for manufacturing, waste from the steel mills, and buildings erected on the floodplains contributed to poor air and water quality, periodic flooding of the city, and hillsides, ravines, and valleys filled with contaminated soils and waste.[21] By the mid-twentieth century, water quality had been greatly improved from 100 years earlier, although deaths by typhoid still occasionally occurred. Water filtration and sewage treatment plants along with an extensive network of storm and sewer pipes made a difference. The rivers were less polluted. Nevertheless, runoff from nearby factory sites still posed a threat to health and water quality. A combination of state and federal legislation in the 1970s and a major collapse of the steel industry in the 1980s greatly improved conditions along and in the rivers. Too-frequent flooding, which often left the low-lying downtown inundated, was brought under control by dams that better controlled the rivers' flow in the spring.

After World War II, poor air quality was still a major problem. During the day, the city would often be enveloped by smog created by the smoke and particulate matter emitted by the steel mills; in the evenings, home heating systems did similar damage. In 1940, 81 percent (142,000) of the city's dwelling units burned coal.[22] Respiratory problems among the residents were prevalent, clothing hung outside to dry would be covered with soot, and, on the worst days, the street lights would turn on during the daylight hours in response to the man-made darkness. Public health advocates, housewives, and civic boosters lobbied the state and local governments for a solution. In response, the city mandated the use of clean, smokeless fuels for heating and this had a significant and positive impact on air quality. Later, the conversion to natural gas and electricity for heating and cooking was also beneficial. The factories and mines were still a problem with the labor unions worried that pollution controls and a shift away from coal would mean a loss of jobs. As the economy transitioned away from manufacturing and the region's steel industry declined, these controls became less of a potential burden and were widely implemented.

Pittsburgh's industry and a lack of land-use controls also contributed to soil contamination and the degradation of hillsides and valleys. The closure of the mills in the 1980s left behind brownfield sites that had to be remediated before new development could take place. The production of steel had generated large quantities of slag and this slag had been dumped in ravines and valleys, killing plant and animal life and polluting streams. The postwar expansion of automobile ownership led to the construction of boulevards and highways within the city, a number of which were cut into hillsides and placed along the water's edge. Environmental laws subsequently reduced soil pollution and provided the legal and financial basis for cleaning up the brownfield sites. And, with the shrinkage of the steel industry and resultant decline of the city's resident population, the building of highways ceased as well.

Numerous Federal environmental protections were initiated in the 1970s in response to the severe air and water pollution still emanating from oil refineries, chemical plants, and tanneries. During that decade, the National Environmental Protection Act was passed, clean air and clean water legislation was strengthened, and later, in 1980, Congress established a program to remediate contaminated industrial sites. As a result, not only was pollution from factories and farms abated, but encroachments on wetlands, species habitats, and water bodies became more difficult due to stricter regulations. In New York City, a city government-led effort in the 1970s to fill in the shore of the Hudson River as part of a highway improvement project was blocked by environmental legislation and the courts.[23] Before this time, there would have been no strong legal basis available to prevent this from happening.

Despite these laws and the rise of an environmental movement, an informed citizen can still read of power plants polluting the air with toxins, barges spilling the oil being delivered to city distributors, hillsides collapsing due to excessive runoff from impermeable surfaces, and rivers and lakes being polluted when heavy storms wash chemicals and debris from the land. Take, for example, an event that occurred in New York City in 2005, when water build-up behind a 75-foot-high stone retaining wall caused its collapse and deposited a 150-foot-wide pile of soils, trees, and boulders on a major roadway.[24] And, even with the introduction of catalytic converters and cleaner grade gasoline, air pollution from trucks and automobiles continues to be an environmental and public health problem.

As might be expected, environmental harm has not been confined to the older, industrial cities. (See table 3.1.) Phoenix, a city lacking in the heavy manufacturing associated with the environmental problems of places like Providence, Buffalo, and Pittsburgh, is a good example. Phoenix, San Diego, Dallas, Orlando, and other cities in the south and west of the country are not known for goods production, but for business and financial services, medical care, technology, education, and tourism; that is, economic activities that ostensibly do not pose obvious environmental risks. Yet, these cities still have industries causing environmental damage. Many had and continue to have manufacturing districts and railroad yards and ports that are not quite as "clean" (environmentally speaking) as banking. These cities were not established after "dirty" manufacturing had closed or moved offshore. They are not pristinely post-industrial. Moreover, many contemporary manufacturing processes (for example, the production of semiconductors in San Jose, California) have serious environmental consequences.[25] And we should not forget the airports, parking lots, and limited-access highways whose environmental impacts have to be constantly mitigated.

The ethnographer Andrew Ross has labeled Phoenix the country's least sustainable city.[26] His claim might be borderline hyperbolic, but his case for the environmental inadvisability and destructiveness of

Table 3.1 **Least Green Cities in the United States, 2015**

Rank	City
1	Baton Rouge, LA
2	Gilbert, AZ
3	Indianapolis, IN
4	Louisville, KY
5	Hialeah, FL
6	Chandler, AZ
7	Jacksonville, FL
8	Glendale, AZ
9	Bakersfield, CA
10	Memphis, TN

Note: Based on a scale developed from measures of environmental quality, "green" transportation, energy sources, and local policies.
Source: Richard Bernardo, "2015's Greenest Cities in America," WalletHub website, accessed April 7, 2016, www.wallethub.com/edu/most-least-green-cities/16246.

Phoenix is convincing. The city is located in a region with fewer than 8 inches of annual rainfall and with aquifers insufficient for supporting a city of 1.6 million people and a metropolitan area of 4.2 million people that sprawls over 1,000 square miles. Consequently, water has to be imported from 335 miles away. In addition, Phoenix is located in one of the hottest areas of the country and requires extensive use of energy-consuming air conditioning to make it livable in the summer. It also "imports almost everything its residents consume: water, energy, manufactured goods, and perishable products," thus giving it an extensive ecological footprint.[27] Being a low-density city with little mass transit, its residents rely heavily on the automobile. This, and the fact that it is also in a geological basin, has led to high levels of dust pollution and unhealthy ozone levels. With its economy heavily dependent on new construction and the real estate industry assertive in its push for further expansion, usually in the form of land-hungry residential subdivisions, farmland continues to shrink as threats to fragile ecological areas persist.

Phoenix's environmental destructiveness is due not just to its ill-advised location, its sprawl, and the ubiquity of the automobile, it is also related to its industry. Rapid growth and the accompanying construction of new homes, office buildings, schools, shopping malls, and highways mean an urban periphery with numerous sand and gravel operators and asphalt plants that pollute the environment. The region is also burdened by toxic waste from uranium mines, chemical dumping from factories, numerous hazardous waste facilities, and landfills for disposing of construction debris and the detritus of consumption. A pro-growth local culture, moreover, restricts regulation of sprawl and blocks outright prohibition by the government of such destructive activities.

The toxicity of Phoenix falls heavily on the poor, and particularly African-Americans, Latinos, and Native Americans who are segregated from the affluent, white population. Those who are marginalized tend to live closer to polluting factories and other environmentally suspect land uses and, as a consequence, suffer health problems. Affluent, white residents are able to isolate themselves from environmental harm. Ross writes that "there is nothing sustainable in the long run about one population living the green American dream while, across town, another is still trapped in poverty and pestilence."[28]

Using an even more alarmist tone, the urban critic Mike Davis has documented the ecological conundrum that is Los Angeles. In its hills wander cougars (mountain lions) that prey on pets and, rarely, on humans. Snakes wash up on the beaches. Storms cause flooding and firestorms race through the dry lands of Malibu (an ultra-affluent community), destroying homes and threatening the lives of animals and the people who live there. Earthquakes are unavoidable given that the city straddles a major geological fault. The earthquake in 1994 left 72 people dead, 12,000 injured, and 25,000 homeless. The total damages were estimated at $42 billion with 437,000 homes in need of repair. Davis concludes that "Los Angeles has deliberately put itself in harm's way. For generations, market-driven urbanization has transgressed environmental common sense."[29]

Urban Metabolism and Sprawl

These descriptions of a city's environmental impacts are informed by what scientists label a city's metabolism; that is, the flows of nutrients, energy, raw materials, consumer and producer products, air and water emissions, and residues (for example, human waste) into and out of the city. Such material flows are essential for human life. The focus is food plants and fossil fuels that have to be brought into the city and the waste that is produced and has to flow outward. In the mid-1960s, Abel Wolman estimated that the average city dweller, on a daily basis, used 150 gallons of water, consumed four pounds of food, and required 19 pounds of fossil fuel to heat and air condition his home and business and travel to and from work and take other trips.[30] Each day, these materials had to either flow into the city or be produced there. Their use and consumption, in turn, generated human waste, used packaging, uneaten food, and the various by-products of fuel consumption—all of which had to be discarded.

Studies of urban metabolism build on this basic observation. Clearly, the material flows are going to be much larger for bigger cities and the problems of managing them much more complex. The New York City metropolis, for example, uses more energy and water and dispenses of more solid waste than the smaller metropolitan area of Los Angeles.[31] These concentrated and large material flows, moreover, have destructive consequences that include "altered ground water effects, exhaustion of local materials, accumulation of toxic materials, summer heat islands, and irregular accumulation of nutrients."[32] Available evidence points to buildings and transportation as being responsible for the largest amount of total energy consumption. Numerous, multi-story buildings and large-scale transportation systems are what define a city in contemporary times, thus making cities much more energy dependent than rural areas. Evidence further suggests that the metabolism of cities is intensifying, particularly as more and more people move into them.[33]

Stepping back from these examples presents us with another picture of environmental destructiveness. As cities grew in population size and number, they increased in density and expanded across land once considered suburban or even rural. Until European colonization in the seventeenth century, much of this land was inhabited by Native Americans and used for small-scale hunting, fishing, subsistence farming, and the gathering of wood and plants. By the mid-nineteenth century, land adjacent to growing cities was often being farmed with single homes and small factories scattered about and small towns here and there. By the mid-twentieth century, mass suburbanization was underway.

Land absorption is one of the most destructive consequences of urbanization. Currently, more than 380 metropolitan areas occupy approximately 3 percent of the land of the United States. Much of the remaining land, however, is unbuildable. Even in exurban areas just beyond the suburban fringe, only a small proportion of the land (about 16 percent) is developable. Developable land is most abundant in rural areas, but that is not where people are interested in living. These lands are unprotected by governmental designation and thus open to many different uses. Because of soil conditions and extreme slopes, however, building on them is likely to be difficult and costly.[34]

Until widespread automobile ownership and usage, the land absorbed into cities roughly matched the growth of the population. Contemporary modes of transportation—commuter rail lines, streetcars—were not designed to serve low-density development and their effect was to concentrate the population geographically, not to disperse it. And even though people had automobiles as far back as the 1900s, they were expensive and thus uncommon with too few highways existing to make it easy to commute from small towns in the metropolitan periphery to the central city.[35] With the surge in automobile ownership and highway construction after the 1940s, land was absorbed at a much greater rate than before. Population growth and land absorption became disengaged. Between 1950 and 1980, the number of people living in metropolitan areas went from 84.5 million to 169.4 million and the amount of land in urban use went from just less than 20 million acres to just over 50 million acres.[36] Of particular note is that "U.S. cities have been built on the most fertile soils." Cities have accounted for a 1.6 percent annual decline in the net primary productivity of land as measured by plant growth. This has been offset by a 1.8 percent annual increase from the expansion of farmland.[37] Adding to the demographic pressure, households are becoming smaller (with more people living alone) and homes are becoming bigger. The average household size in the United States has fallen from 3.37 persons in 1950 to 2.59 persons in 2010, while the average home size went from 1,730 square feet in 1983 to 2,600 square feet in 2013.[38] These factors amplify the impact of urbanization on the land and, even though this land is not necessarily despoiled, much of it had once been used for agriculture and its loss is significant. Development, moreover, often brings soil erosion along with water and air pollution.

Over the past four or five decades, this concern for the encroachment on productive farmland and ecological areas has been debated around the issue of sprawl. The postwar suburban development that came to define the US metropolis was one of low-density communities organized around the automobile, with the automobile geographically dispersing rather than geographically concentrating homes, businesses, and public services such as schools, parks, and playgrounds. Sprawl appears when "rapid, unplanned, or at least uncoordinated, scattered, low-density, automobile-dependent growth [occurs] at the edge or in the urban periphery." It is the consequence of a complex array of conditions: a growing population, rising incomes, greater automobile use and highway construction, falling commuter costs, federal policies, weak development controls, and popular attitudes related to race and poverty. One observer described sprawl as "the new kind of postindustrial ugliness that was overspreading the landscape."[39]

A 2014 study of sprawl in the country's metropolitan areas used four factors to measure its existence: the density of development, the mix of land uses, the spatial juxtaposition of people and jobs, and the connectivity of streets and their accessibility to pedestrians.[40] Using these four factors, the authors created an index that ranked the metropolitan areas in terms of how compact and interconnected they were. At the top of the list were New York City, San Francisco, Atlantic City (NJ), Santa Barbara (CA), and Champaign/Urbana (IL). At the bottom of the list were Hickory (NC), Atlanta, Clarksville (TN), Prescott (AZ), and Nashville. The study then noted that people in more compact and connected metropolitan areas have greater economic mobility, spend less on the combined costs of housing and transportation (even though they spend more on housing), and have more transportation options. They also walk more and are more likely to use mass transit and to be less obese and live healthier, safer, and longer lives.

Critics of sprawl—those who champion dense and compact cities—

have enumerated a long list of problems. Sprawl causes excessive travel due to the spreading out of destinations, is aesthetically displeasing, contributes to a boring and conformist lifestyle, diverts much-needed financing from the cities (particularly for mass transit), and leads to racial and economic segregation as minorities are blocked from entering the suburbs and jobs are located far from those who most need them.[41] One of the major criticisms is that sprawl is environmentally destructive: it encourages more automobile use (increasing greenhouse gas emissions) and highway construction (taking up more land), makes inefficient use of water, sewage, and electrical infrastructure, and leads to water-intensive lawns and plantings as well as impermeable surfaces (especially highways and large parking lots) that cause erosion and flooding and contribute to water pollution. Additionally, the spreading out of homes impinges on farmland as well as wildlife and plant habitats. As one critic has commented, "Few can argue that low-density development does not increase auto emissions, water use, pollution, trash, loss of species habitat, and energy consumption."[42]

The defense of sprawl is two-fold. First, people want to live in suburban environments as indicated by the demand for suburban homes and the relatively slow growth in the population of central cities. In fact, many of sprawl's defenders consider it "the preferred settlement pattern everywhere in the world."[43] The truth of this claim is debatable. Despite much concern about central city population decline in the decades after World War II, the central city population went from 53.7 million in 1950 to 77.8 million in 1990 and to 114.1 million in 2010. Admittedly, much of this increase had to do with towns growing into cities rather than older cities expanding; that is, the growth is partly a statistical phenomenon. Second, low-density development can be—much in the same way that cities are said to have the potential to be—environmentally sustainable, but only with the addition of more mass transit options, alternative energy sources, fuel-efficient automobiles, more compact development, and stewardship of public lands. In short, only if they become more like cities.

Whether sprawling or not, suburbs pose threats to air, water, and soil quality as well as being major uses of nonrenewable energy. In fact, the postwar suburbs were a major factor in the emergence of the environmental movement. Absent environmental laws pertaining to low-density construction, animal habitats were disrupted, forests were denuded, wetlands were filled, and soils were damaged so that large, flat expanses of land could be prepared for street after street of single-family houses. Insensitivity toward nature on the city's periphery, less than the threats to wilderness, led to the passage of protective legislation.[44] Even in the 2010s—and despite the spate of environmental regulations that monitor and govern manufacturing processes, transportation, construction, land development, and ecological remediation—cities and suburbs still coexist with nature in ways that are more threatening than respectful.

One of the destructive sides of cities is captured in the notion of the ecological footprint.[45] This footprint measures the land and water area required to absorb the city's waste and provide it with the energy and the materials (for example, corn, gravel, and water) that it needs to function and for residents to enjoy a desirable quality of life. Or, to use a more evocative definition, the consumption footprint is the sum of "the cropland, grazing land, forest land, fishing ground, built-up land, and carbon uptake land . . . . required to produce the food, fibre, and timber [a city] consumes, and to absorb the carbon dioxide waste it generates." The emphasis is on the flow of energy and matter into and out of the city with the measure capturing the impact of a city on nearby and far-flung ecosystems. And although land absorption is "only a small fraction of their environmental impact," land is still central to this way of thinking about the environmental effects of cities.[46] Ideally, and to be environmental sustainable, the provision of resources and absorption of wastes should not deplete the capital stocks within the footprint or undermine ecological systems. One of the overarching questions is the extent to which a city runs an ecological deficit or sustainability gap through its use of nonrenewable resources.

Not surprisingly, cities have extensive ecological footprints. They draw resources from around the world and send waste such as obsolete electronic products well beyond their political boundaries. Consequently, cities use up much more of the planet than they occupy; they exceed the carrying capacity of the land on which they sit. For cities in advanced economies, food comes from across the country and beyond their borders, electronics and carbon-based fuels are imported, clothing is produced in low-wage economies and transported in container ships to local markets, and water (as in Phoenix, Las Vegas, and Los Angeles) is brought from hundreds of miles away. People might live in local places, but those who live in cities can do so only because businesses, transportation systems, and financial arrangements pull resources from elsewhere. Not to do so would mean living at subsistence levels, depriving most middle-class people of their way of life.

Well known is that the United States is far from exemplary when it comes to such environmental matters. Of 22 countries that include

Pakistan, Sweden, and Australia, the United States in 2002 had the largest ecological footprint and its ecological deficit was the highest. In fact, it was 13 times the average for the group. A study of the resource flows of 27 megacities with populations exceeding 10 million in 2010 ranked New York City as the highest in total energy consumption, water use, and solid waste disposal. Los Angeles ranked fifth.[47] As regards the ecological footprints of US cities, researchers for the Global Footprint Network put those of Seattle, Washington, DC, and Minneapolis above that of New York City. Using global hectares consumed per capita as its measure, all of the former metropolitan areas ranked over the US average, while New York City ranked below it. A more detailed analysis for San Francisco found that the three biggest contributors to its consumption ecological footprint were food and beverages, transport, and restaurants and hotels.[48]

Big cities have large ecological footprints. At the same time, they are less environmentally destructive than smaller cities while cities overall are less environmentally destructive than suburbs and rural settlements. (See table 3.2.) The total household carbon footprint of large cities is 80 percent of the large suburbs. While cities have extensive ecological footprints, then, they are also more efficient users of energy, a key element of environmental sustainability. This assessment finds additional support in data on energy use in cities. For 2007, greenhouse gas emission for the United States was estimated to be 23.6 metric tons of carbon dioxide equivalence per capita. For cities, this number was lower: 21.5 for Denver, 18.3 for Minneapolis, 13.7 for Seattle, and 10.5 for New York City.[49] Cities might be environmentally suspect, but they are less so than other forms of human settlement.

Table 3.2 Carbon Footprints by Settlement Type

Settlement Type	Total Household Carbon Footprint
Large cities	41.8
Mid-size cities	45.1
Small cities	46.6
Rural remote areas	47.6
Small suburbs	50.0
Mid-size suburbs	51.0
Large suburbs	53.1

Source: Christopher Jones and Daniel M. Kammen, "Spatial Distributions of U.S. Household Carbon Footprints Reveal Suburbanization Undermines Greenhouse Gas Benefits of Urban Population Density," *Environmental Science & Technology* 48, no. 12 (2013):895–902. See Table 1, p. 899.

Undeniable is that cities are environmentally disruptive even if not as destructive as they were during the period of industrialization and before extensive environmental regulations. When humans are brought together in such large and dense human settlements, nonrenewable resources are consumed, land is absorbed, and ecosystems are displaced, disturbed, and, at times, decimated. That cities might be less environmentally destructive than other forms of human settlement and that they might have great potential to be sustainable is pivotal to this environmental contradiction.

Quest for Sustainability

Even though much evidence points to the environmental destructiveness of cities, it is generally claimed that cities are our most sustainable settlement option. The argument travels along three paths: density, technology, and the inherent superiority of cities over their main competitors—the suburbs. Each of these concerns is premised on the belief that the spatial form of the city is critical for being sustainable. This premise, however, leaves aside important issues such as the energy sources utilized to heat and cool buildings, operate transit systems, and fuel automobiles.

Density, the first of these concerns, refers to the compact spatial clustering of people, activities, and structures that enables energy use to be minimized. The journalist Alex Marshall is clear about this: "increased population will help mass transit work. It will help neighborhood business districts and downtown. It will further push the center of gravity away from the suburban malls and subdivisions, and toward the center city areas and train stops."[50]

With buildings sharing walls, people traveling together in buses and subway cars, and public services achieving economies of scale so as to be more efficient, energy is saved relative to what occurs in less dense and smaller places. Energy is the key to sustainability because of the current reliance on nonrenewable energy sources (coal, gas) and the extensive environmental costs of extracting, shipping, processing, and consuming these forms of energy. As one advocate has noted, "Because urbanites share transportation and land in close proximity, they use far fewer resources per person and destroy far less wilderness" than those who live outside cities.[51] Density thereby intersects with different modes of transportation, building design, land absorption, and urban form to establish a basis on which to pursue policies that reduce energy use and minimize the impacts that cities have on ecological systems.

Transportation is another critical element. Because of their density—the concentration of travelers and the proximity of destinations—cities can support subways, buses, light rail, and trolleys and services such as taxis, commuter vans, and automobile-sharing. Since 2000, Camden (NJ), Charlotte (NC), Houston, Little Rock, Tacoma, and Tampa among others have added light rail or street car systems to their transportation options. This reduces automobile usage. People travel more efficiently, less fuel is used, and less land has to be devoted to parking lots and garages. Mass transit also reduces greenhouse gas emissions, although automobile usage still predominates in all US cities except for New York. And, because destinations are closer together, people are more likely to walk or even bike, thereby shifting to renewable energy sources. Consequently, many city governments in the United States have passed laws to discourage an overabundance of parking spaces, considered reducing automobile access to the central business district by assessing tolls for doing so, and encouraged greater bicycle use and automobile sharing.[52]

Buildings are also major energy users and the form they take in dense, urban settings holds out another possibility for increasing sustainability. Buildings "consume 71 percent of the nation's electric power and 39 percent of all power, and create 39 percent of the nation's CO2 emissions."[53] Apartment buildings use less energy per household than single-family, detached houses or garden apartments. They share exterior walls, thus diminishing energy loss, and have less exterior surface per occupant than detached, single-family homes. Tall office buildings have similar environmental qualities. Not only are they already relatively energy-efficient when compared to office parks (where the buildings are low-rise and separated from each other), but they occupy less land per worker. And, the larger size of office towers and apartment buildings in cities compared to suburbs means that they house more activities per acre. These buildings can be made even more energy-efficient and sustainable through recycling of waste during construction, the use of low-energy materials such as wood, green roofs that reduce heat absorption and capture rain water, gray water systems that recirculate water rather than simply disposing of it, climate-sensitive exterior walls, regulated interior lighting systems, air recycling, and shared heating arrangements. Both Seattle and Houston have established initiatives to encourage sustainable building projects, with Houston focusing on their energy efficiency in order to counteract the heavy use of air conditioning. Berkeley (CA), among many other cities, has a program to enable homeowners to convert to solar energy.[54] Many of these sustainable building technologies are made financially feasible because of the large scale of buildings; converting a boiler to cleaner fuels in an apartment building has an impact across a greater number of people than converting furnaces in single-family detached houses.

To state the obvious, dense cities also use less land than less dense places. Generally, cities do not encroach as extensively on farms, woodlands, and prairies as, for example, sprawling suburbs. Leaving woodlands, nearby farms, meadows, and streams unencumbered with development allows animals, birds, and plants to thrive with all the benefits of carbon sequestration, absorption of stormwater, and protection of biodiversity. And while dense cities are more likely to set aside land for parks to be used for human recreation, rather than left unspoiled by human use, even these recreational areas have positive environmental benefits. In a number of major cities such as Philadelphia or St. Louis, the major parks are large enough to support not just humans but significant numbers of animals, birds, and insects not to mention plant life. Many cities have begun to design green spaces to control stormwater runoff, have daylighted streams (that is, restored streams to their natural state above ground), encouraged the use of permeable surfaces, softened their waterfronts, and developed protections for wild animals and aquatic creatures as well as local plant life. Philadelphia is a good example of these latter efforts, having embarked on a citywide stormwater management plan.[55]

A city's contribution to lessening land absorption has much to do with urban form; that is, the way buildings and structures are arranged in a city. This encompasses not just compactness but also the juxtaposition of different types of land-using activities. The argument is that sustainability rests on how cities are spatially organized. Simply being dense is insufficient. Much as a building benefits from being properly oriented in terms of sun and wind, a city benefits environmentally when it is sited to minimize its impact on ecologically sensitive areas and when the different activities it contains are arranged so as to minimize movement between them. One comparative analysis of urban form indicated that 5.25 tons of carbon emissions were generated annually by each household at average suburban densities whereas at average urban densities each household generated 1.29 tons. Moreover, the energy consumed by the average urban household was one-half of that consumed by the average suburban household.[56]

This concern with compact urban form is most apparent when governments establish growth boundaries around their cities. These boundaries constrict the spread of development in order to preserve farmland and wildlife areas and achieve economies of scale in infrastructure use through higher densities. Essentially, land-use regulations are used to block or severely curtail development on the city's urban fringe, thereby directing it inward to already built-up areas and encouraging, as an additional environmental benefit, the adaptive reuse and upgrading of existing buildings and structures. Portland (OR) is the country's prime example of a growth boundary.[57]

Two of this argument's central concerns are mixed-use areas and transit-oriented developments. The first is a reaction to the segregation of land uses that was championed by early twentieth-century city planning. In response to the chaos of the industrial city that led to homes being built adjacent to noisy and polluting factories and factories encroaching on emerging office districts, the early city planners argued for and achieved a separation of land uses. Residential, commercial, and industrial activities would occupy different parts of the city. This minimized the undesirable impacts (known as negative externalities) of one land use on another. The noise of a metal fabrication factory would be too far away from residential areas to disturb anyone's sleep and the smells of a poultry processing plant would not drift into a retail district. But while this protected people from noise, smoke, and the fumes of idling trucks and locomotives and, as a consequence, preserved their property values, it also maximized the amount of travel among and between these activities and places. Doctors' offices, hardware stores, movie theaters, and homes were all in their distinct and separate places. People no longer went a few streets away to work or walked mere blocks to the grocery store. Now, they had to travel by bus or automobile to do their daily tasks.

By the late twentieth century, large cities no longer harbored noxious industries and the shipping that served them. Consequently, many planners, developers, and elected officials began to recognize the value of mixing rather than separating land uses. Bringing retail, residences, and work spaces together came to be seen as the best way to encourage more lively public spaces. People would live close to their place of employment, while restaurants, retail shops, specialty food stores, cinemas, and clubs would be within walking distance of their homes.

Unlike the industrial city, the city would be organized around consumption rather than production. Doing so would decrease automobile travel (saving energy and diminishing pollution) and make life more attractive simply by being more urban. Such mixed-use areas benefit greatly from substantial densities. When densities are low, stores cannot remain open, public libraries are less viable, and restaurants are vulnerable to financial failure.[58] With high densities, they thrive.

This idea of mixed-use areas rests on the energy-saving benefits of mass transit. The two come together in the form of transit-oriented developments (TODs) that combine mixed-use areas with mass transit stops.[59] The idea is that higher residential densities around subway stops, light rail stations, or bus-rapid transit stations will encourage people to walk more (since mass transit is nearby) and provide the requisite density to support the transit node. The mix of uses will decrease the amount of travel for those who work close by, particularly when it includes offices and stores that provide jobs. Others will have to commute to the central office area or to transit nodes where they are employed, but can do so using the mass transit available to them in their neighborhood. With mixed uses, including a critical mass of housing, even commuters will not have to drive to the stations. Over time, the existence of these stations will spur even greater densities as businesses and people locate to where life is more convenient and urban amenities more accessible. Higher densities will attract developers and investors as well. All of this is a considerable improvement over the sprawl of the suburbs where all trips require an automobile.

The second path along which the argument for the inherent sustainability of cities travels is that of technologies. Density enables a variety of technologies to be deployed at a scale that saves energy, reduces land absorption, and minimizes environmental burdens. Mass transit is more feasible, thermal heating systems can be shared, water runoff can be controlled through green roofs and permeable surfaces, street lighting can be regulated with sensing devices, and car-sharing arrangements can be made profitable. In addition, local governments can promote bicycle and pedestrian alternatives and emphasize "access by proximity." Cities also enable various energy-saving building technologies to be widely instituted. These include gray water systems that recycle the water from wash basins and clothes washers, glass curtain walls on office buildings that include electronically controlled louvers that modulate heat loss and absorption depending on the outside temperature, and greater use of solar panels.[60] Waste recycling processes minimize the use of landfills, while technologies for capturing the reusable material from construction and demolition further contribute to sustainability.

Local governments are particularly important in this regard since regulatory powers and sensitivity to broad social consequences, rather than narrow profitability, enable them to adopt emerging technologies such as hybrid vehicles and solar power. They can test these innovations and serve as an example for the private sector.[61] In addition, government regulations are essential for activities such as recycling and for shifting travelers to mass transit by limiting the parking spaces in office districts. One of the most important of these technologies is mass transit, either buses or subways or light rail, that are meant to encourage people to shift from the automobile to more energy-efficient transit modes, thereby reducing air pollution. Governments are also central to greening streets, removing unnecessary pavement, planting trees, and regulating water usage.[62] In all these instances, the density of the city and the size of the market makes these technologies feasible and, in many instances, profitable.

Los Angeles is a good example. Once known as one of the country's most sprawling, automobile-dependent cities, a city without an urban core, and an ecological disaster (due partly to its lack of nearby water sources), it is undergoing significant change to a more sustainable city. The city government has developed a sustainability plan to eliminate coal use, divert 90 percent of waste from landfills, extend rail and bus rapid transit service, and "green" parts of the concrete-channeled Los Angeles River.[63] It has also hired a chief sustainability officer and reduced its own water use. Decades back, it converted to low-flow toilets throughout the city. New housing is being built on already-developed land, thus increasing density and creating the potential for a more urban environment.

The third path along which the sustainability argument travels involves a comparison between cities and suburbs. It sets large, dense cities in contrast with the major form of development in the United States—low-density suburbanization. Cities are considered sustainable in comparison to the suburbs.[64] Automobile-dependent, land-wasting, ecology-destroying, and consumption-fixated, suburbs use up resources and energy much beyond that which occurs in cities. In cities, people walk more, have smaller homes, and use less fuel for heat in the winter and air conditioning in the summer. Even compared to small towns and villages, suburbs are vilified as environmentally harmful. Just as the suburb was meant to combine the countryside with the city, providing not only open space for private use but also access to stores and entertainment and jobs, the city has come to be seen as another variation on this merging of two worlds that began in the early twentieth century with Ebenezer Howard's garden cities.[65] In this instance, however, nature has been filtered through technology rather than left in its ostensible original state or reimagined as well-manicured lawns.

All forms of human settlement have the potential to be more sustainable, but this potential seems greater for cities than for low-density suburbs. The "ideal" city is the sustainable city populated by individuals and technologies that minimize energy use and discourage further encroachment on the natural environment. Sustainability will be achieved through density and through the potential for shared activities and economies of scale that density promises. Moreover, governments will play a central role in passing laws and subsidizing businesses and households to be environmentally responsible.[66]

Key to the inclination to sustainability is the notion of sharing. Consider the seemingly extreme claim that "nowhere in human culture is the centrality of collaboration and sharing more obvious than in the city."[67] Despite the concentration of private wealth, and the selfishness that it implies, cities are considered places where people are literally compelled to share buses, sidewalks, taxis, and apartment houses and continually brought into contact with each other in ways that engender mutual recognition and even assistance. Constant interaction serves as a basis for sharing and it enables goods and services, infrastructure, and places to be heavily used, thereby avoiding the environmental costs of individual ownership and use.

While heightening densities to decrease land absorption, expanding mass transit use, and recycling waste and water can contribute to making the city more sustainable, they have only a minor impact on reducing its ecological footprint. More critical in importance is reducing energy use by buildings and, at the top of the list, shifting to renewable energy sources. Seattle's reputation as a sustainable city rests on the fact that 90 percent of its electricity comes from hydropower and the remaining 10 percent is offset by composting, methane recapture, and biodiesel fuels.

As for reducing a city's ecological footprint, the primary initiative in this regard is making office towers, apartment buildings, retail stores, and factories much more energy-efficient, including the creation of district energy systems. What receives more publicity, though, is the local sourcing of material flows into the city, specifically of food. Through the use of farmers' or "green" markets and public pressure to encourage households and restaurants to buy fruit and vegetables, chicken and meat, herbs, and fish from nearby producers, the intent is to reduce the distance traveled by food as it enters the city.[68] The ecological footprint can also be reduced by considering the geographical origins of the inputs to nonfood businesses (such as printing companies) and local government. In addition, local economic development agencies are working with planning agencies to set aside land for "spatially integrated" business clusters so as to minimize the reliance on transportation of materials in and products out, similar to considerations involving the movement of workers between their homes and their jobs.

A number of cities have earned a deserved reputation for efforts to make themselves more sustainable.[69] (See table 3.3.) Most renowned in the United States is Portland, Oregon. Not only does it have a famous growth boundary meant to forestall sprawl by concentrating development toward the center of the city, but in 2013 approximately one-half of its energy came from renewable sources. Second on a recent ranking was San Francisco with 80 percent of its waste being recycled or composted and over 700 buildings certified as "green." Other cities such as Seattle, Austin, New York City, and Grand Rapids have embarked on numerous efforts to increase recycling, reduce greenhouse gas emissions, create more parkland and expand the tree coverage, provide bikeways, and replace gas-powered government trucks, vans, buses, and cars with alternative fuel vehicles. These and other cities have also developed climate change plans to identify ways to reduce greenhouse gas emissions

Table 3.3 **Most Sustainable Cities in the United States, 2013**

Ranking	City
1	Portland, OR
2	San Francisco
3	Seattle
4	Minneapolis
5	Austin
6	Eugene, OR
7	New York City
8	Salt Lake City
9	Grand Rapids, MI
10	Philadelphia

Source: John Light, "Cities Leading the Way in Sustainability," posted 2013, accessed July 1. 2015, www.billmoyers.com/content/12_cities_leading_the_way_in_sustainability. See also Elizabeth Svoboda, "American's 50 Greenest Cities," *Popular Science*, posted 2008, accessed April 7, 2016, www.popsci.com/environment/article/2008-02/americas-50-greenest-cities.

and protect against rising sea levels, the increasing severity of storms, and an intensifying heat island effect.[70]

Despite these "success stories," it is difficult to assess definitively the impact of efforts to make cities more ecologically viable and reduce their ecological footprints. Certainly these initiatives are admirable and each, in their own way, moves cities closer to the loosely defined state of sustainability. That similar activities might occur in the suburbs, possibly not as efficiently or with as much of an impact, is a reminder that we live in a world of relatives rather than absolutes. It should also remind us that only a minority of the country's residents live in the kind of large and dense cities—New York, Chicago, San Francisco, Seattle—on which this urban sustainability argument is based. Consequently, the claim that "If the future is going to be green, then it must be more urban" should be read skeptically.[71]

Growth, Decline, Resilience

The quest for sustainability seemingly ignores the very problem that it hopes to solve; that is, that the city is an often destructive intrusion and little can be accomplished short of shrinking the population, dismantling the technologies on which the country's standard of living is based and to which many aspire, and shifting in a major way to renewable energy. It would also require a return to settlement patterns that have not existed for some time. Even after 100 or so years of enlightened concern about the impact of humans on the natural environment, humans continue to absorb that environment into the city in ways that diminish animals, plants, fish, soils, insects, land forms, and bodies of water. The justification is a hubristic rationale that sets the material world in service to humans. Poorly recognized is both the inseparability of human culture from nature and the interdependencies without which human life would be impossible. Ignoring those interdependencies is at the root of the city's destructiveness.

Yet, it has also been obvious to humans for centuries that these interdependencies have to be respected if humans wish to live in certain ways. Farmers, early on, discovered and then adapted to the need to replenish the soils on which they grew their crops. City dwellers learned quickly that one might want to separate privies from wells. Even before the emergence of governmental regulations in the late nineteenth century, city residents knew enough to modulate their relation to nature, even if they were generally ignorant of the scientific explanations for their difficulties. The Industrial Revolution brought these destructive tendencies into sharper focus and made their economic and health consequences politically unavoidable. Since that time, efforts at minimizing the environmental burden of cities and, more recently, attempting to make them more sustainable have continued to multiply, and with positive effects. No matter how sustainable cities become, however, they can never be made sympathetic with nature. They are an imposition and will continue to do damage.

The contradiction between environmental sustainability and destructiveness is exacerbated as cities increase in size and density and as growth races ahead of the ability of local governments to mitigate its negative consequences through public health regulations, land use guidelines, and public infrastructures. That said, neither growing nor declining cities have a monopoly on managing this tension. Growing cities might absorb more land, encumber water and sewage systems and roadways, and add pollutants to the air and water, but declining cities produce contaminated and abandoned sites where manufacturing and shipping once thrived. Although the former displace animal, fish, and plant habitats, the latter often enable them to return. In Detroit, as buildings were demolished and block after block was left empty, pheasants, rabbits, hawks, and other wildlife reappeared along with native plants.[72] The problem for governments in growing cities is one of keeping pace with growth—minimizing its environmental impact and responding with the necessary technologies for serving new residents. Opportunities also exist for creating mixed-use, transit-oriented developments, encouraging "green" construction practices and buildings, and mandating the use of native plants. For governments in declining cities, the problem is one of reimagining the city as smaller and less dense and treating this as an opportunity for becoming more sustainable. Soils can be remediated, woodlands can be reestablished, formerly buried streams can be reopened, and land can be set aside to absorb rainwater runoff.[73]

Once sustainability has been achieved, and for it to endure, the city has to be made resilient.[74] The state of sustainability has to be maintained and this requires city governments to respond and adapt to more frequent and more severe storms, sea level rise brought about by climate change, and unexpected events such as chemical factory explosions, earthquakes, and acts of terrorism. The ability to respond to natural disasters is a persistent problem. New York City recovered relatively quickly from the flooding and damage of Hurricane Sandy in 2012, but New Orleans has yet to do so from Hurricane Katrina in

2005. Then, there is the slow and unavoidable fact of climate change with coastal cities such as Miami and Galveston highly vulnerable to sea level rise and ignoring it at their peril. Specific responses range from disaster plans and building codes that protect homes from flooding to strengthened sea walls, back-up power supplies, and neighborhood evacuation procedures. Resilience points to the need to adapt to the ever-changing accommodations struck by humans, nature, and technologies.

The relation of this contradiction to wealth and poverty is not so obvious. One might expect that cities with large, wealthy populations and particularly a large middle-class would produce more political support for protecting the environment while generating tax revenues sufficient to develop programs that encourage sustainability. This does, in general, seem to be the case. The cities known for their sustainability initiatives are places like Seattle, New York City, Portland (OR), and Austin where housing prices are high, households are relatively affluent, and local government competent and engaged. At the same time, in certain poor and shrinking cities—Detroit and Cleveland come to mind—lacking these qualities—local foundations, nonprofit advocacy groups, and universities have mounted initiatives to address environmental quality.[75] And in wealthy cities, particularly those that are growing rapidly, such as Phoenix and San Jose, local political and economic elites often sacrifice environmental protection for growth. Moreover, while local governments in cities with large poor populations are also likely to have an anemic tax base with fewer resources to devote to nonessential services—the essential ones being police, fire, and waste management—they do have access to state and federal governmental programs. In wealthier cities, having a strong tax base does not directly translate into an array of programs and regulations directed at environmental sustainability. These are, of course, political decisions.

This brings us to governance and the contribution that public actions make to life in the city. Politics does not affect economies and environments from a position outside of them. Rather, political arrangements are an integral part of their formation and thus of the consequences that cities engender.[76] That political arrangements vacillate between oligarchy and democracy is another contradiction in the life of US cities.

FOUR

Oligarchic, Democratic

Cities would seem to be ideal places for democracy to flourish. Living in close proximity to each other, forced to share services and roads and attend to public health, and wishing to live harmoniously, residents are compelled to negotiate differences regarding taxation and matters of religion, ethnicity, national origin, and lifestyle. The need to resolve the many and often unpredictable frictions that cities engender encourages people to embrace democratic practices that distribute rather than concentrate power, legally guarantee individual and group rights, enable a free press, allow for direct rule, and establish the basis for mutual trust.[1] Added to this is the sense of mastery that people experience as they successfully engage with strangers. With its unpredictability and multiplicity, the city rewards those who are adept at meeting its challenges. In turn, this enriches the ground on which democracy can thrive.[2] A politics of coexistence and tolerance has additional benefits as well; it gives collective meaning to people's lives and instills the responsibilities of citizenship. Those who live in cities have ample reasons to gravitate to forms of governance in which they collectively influence and jointly participate in the decisions that bear upon their daily existence.

Because of their size and complexity, cities are also difficult to govern in a centralized fashion. Governing becomes easier when residents add their knowledge and support to common endeavors. The decentralization of decision-making and the delivery of public services allow governments to adapt to the varied interests of residents, show them respect, and foster civic awareness. Governments enhance their legitimacy and stave off opposition by becoming accountable. In contrast, policies that erode broad political support or disenfranchise and disadvantage particular groups can be costly for elected officials. In such circumstances, mayors and city councils have to redirect resources to rebuilding political support and, in the worst cases, quell civil disturbances. From this perspective as well, the nature of cities seems to encourage and bolster democratic practices.

Historically, democracy has been invigorated by the growth of large cities. "The story of cities," one commentator has written, a bit too glibly, "is the story of democracy."[3] Democratic, representative governments were strengthened when cities in the late nineteenth and early twentieth centuries were faced with widespread poverty, labor and social unrest, and epidemics. These problems required more than the paternalism of local elites and the ministrations of church-based charities. To discipline the city, address the threats of disease and fire, and still dissent, reformers launched more open and democratic practices. Progressive Era reform in the United States spawned numerous mechanisms meant to make government more accountable and accessible. Such arrangements were part of efforts to mitigate the ills and burdens of the industrial city.[4] Citizens were given the opportunity to influence local governmental policy and, with that opportunity, a pathway opened for addressing the less desirable consequences—low wages, excessive hours, child labor—of the capitalist enterprise. As the twentieth century unfolded, governance mechanisms spread at the workplace and in urban neighborhoods. By the twenty-first century, it was widely accepted that citizens should be informed and involved not just at elections, but on a regular basis. Participatory democracy became a common practice of local governments.

Democracy, of course, is not an inevitable outcome of or precondition for urbanization. Cities have also prospered under emperors and dictators. Moscow has grown across decades of nondemocratic rule. During the dictatorship of Joseph Stalin, through the collapse of the Soviet Union in the early 1990s, and now under the authoritarian leadership of Vladimir Putin—from 1939 to 2017—the city has gained population and nearly tripled in size to 12.2 million residents. Washington, DC, a city that lacked self-rule from its founding in 1790 until 1973, nonetheless continued to attract more and more people to live and work there.[5]

That said, many cities in the United States, a representative democracy from its origins as a nation, have been governed against a backdrop of concentrated power and influence that deprives many of the opportunity to participate in the decisions that affect their lives together. This I term *oligarchy*. Political "machines" were formed in the late nineteenth and early twentieth centuries in a number of large cities to control elections, use service provision and government contracting to reward supporters and amass wealth, and block democratic opposition. Such political arrangements took root in Kansas City, Boston, Cincinnati, and New York among many other places. These machines were headed by powerful individuals: Frank Hague in Jersey City (NJ), James Michael Curley in Boston, William Tweed in New York City, and George Cox in Cincinnati. By the late twentieth century, most of the political machines had fallen to reformers. Unsurprisingly, reform was not total. A well-known exception was Major Richard J. Daley in Chicago, who spanned the 1960s. Even in the early twenty-first century, many local governments (for example, Atlanta and Houston) are still run by coalitions of real estate firms, large property owners, and financial institutions that have joined professional politicians to direct governmental resources and regulations to boosting the local economy. These growth coalitions sit on the margins of the democratic process with the concentration of influence limiting and stifling democracy such that the needs and desires of the affluent and well-connected take precedence over those of the average citizen.[6]

Generally, what we find in US cities are democratic practices set within oligarchic frameworks comprising economic, political, and cultural elites. Free elections regularly occur, the press is allowed to speak, and individual and group rights are protected by the courts, but the big decisions of government regarding capital investments, tax policy, schooling, and business incentives are mainly the purview of well-organized and well-funded interests. Even in the realm of electoral politics where democracy is supposed to thrive, political parties and those who make large donations to candidates have a disproportionate influence on who runs for office and the decisions they make there. Lobbyists for business interests, unions, nonprofit institutions (such as universities), and advocacy groups, moreover, are central to the legislative process. Elites, organized interests, and elected officials tend to dominate city governments, particularly as regards policies concerning the local economy and such public services as policing. The influence of elites is disproportionate to their numbers.

Democracy exists not only within and in relation to the local government but also outside of it in neighborhoods, religious associations, grassroots political organizations, workplaces, and civic groups. People assert their democratic rights using a multitude of venues, not just the voting booth or meetings of the city council. In the realm of civil society, they establish governance arrangements independent of, tangential to, and entangled with local governments, the logics of consumer markets, and the strictures of wage-and-salary employment.[7] Through a variety of channels, ordinary citizens strive to manage their lives and their relations to others. They also hope to influence governmental and corporate decisions and, if unsuccessful, mobilize in opposition. Opposition, though, is seldom satisfactory and victories are often limited.

In this chapter, I explore the ways in which cities allow and nurture both undemocratic (oligarchic) and democratic practices. I do this by focusing on governance rather than simply government. There is more to local democracy than what happens within and in reaction to governmental policies. Beyond the politics in which governments are enmeshed are numerous publics that address a range of issues from encouraging dog walkers to clean up after their animals to providing recreational activities for senior citizens and advocating for queer rights. Even when local governments are undemocratic, democracy persists and flourishes in the myriad spaces of the city.

My concern is with how the residents of cities provide for their collective needs and manage their relations with institutions and others unlike themselves, doing so amidst the contradictory impulses of democracy and oligarchy. I begin with civil society—and, specifically, neighborhood-based democracy—and then address the democratic practices of city governments, ending with a brief foray into the extra-local politics of intergovernmental relations. Politics is not only a matter of elections, city council hearings, and mayoral proclamations nor wholly confined to what occurs within the city's boundaries. Rather, the political activities that make a difference extend outside local government and beyond the city itself. And, although a tendency toward oligarchy exists and persists in US cities, it is, like democracy, neither so monolithic nor so entrenched as to be immutable.

Civic Realm

In cities, numerous opportunities exist for people to engage in democratic self-governance and thereby create a civic realm that supports their ways of life.[8] Many of these opportunities, if not the great majority, stem from the needs and desires of residents to engage others in improving their neighborhoods, forming recreational leagues, speaking out on public issues, or simply coming together to worship and provide mutual support. They also encompass the joining of political parties and campaigning for candidates to public office. I will use the term *governance* for the kinds of activities that occur outside, around, and through legally mandated and formal governmental arrangements. Under this rubric are voluntary organizations, neighborhood associations, advocacy groups, religious societies, political parties, and social movements. To begin, I offer a very brief historical comment.

The first European settlers in the United States did not arrive with their local governments preformed and ready to set in place. In New Amsterdam, governance was initially provided by an overseer appointed by a corporation, the Dutch West India Company, who ruled until 1664 when the Dutch surrendered the colony to England and it was renamed New York.[9] In other colonies, the settlers were governed by the church elders around whom they had congregated and with whom they had journeyed across the Atlantic Ocean. Later, in New England, town-wide meetings addressed common concerns and local disagreements. The governing of trading posts by overseers and communities by religious elders, however, hardly qualifies as democratic and, despite being rooted in the community of settlers, colonial governance was usually only open to property-owning men. For the most part, decisions about whether or not to build a wharf, to assign fields for crops or for the grazing of animals, how much food to store for the winter months, and whether to enlarge the fortifications were taken by the settlement's unelected leaders.

As these settlements grew in population, more attention had to be given to where roads and buildings might be placed, how and where to dispose of human waste, and whether and to whom to provide schools. Later, as buildings were located closer and closer together and density increased, issues of fire and water quality became public matters. They required more elaborate responses, initially taking the form of volunteer and subscription fire companies. By the mid-nineteenth century, clear to the city's inhabitants was that more democratic and elaborate local governance arrangements were required.[10] With the onset of industrialization, massive immigration to the cities from Europe, and migration from the countryside, the need for more formal and public responses to housing, water quality, sewage, and street paving become unavoidable. During the Progressive Era of the early twentieth century, various reforms were initiated to create local governments that could regulate development, provide public services, and run democratic elections. From then until today, these governments have become larger and more complex. Nevertheless, they remain, in qualitative terms, much the same as they were a century ago.

That said, one of the most important transformations in governance arrangements has involved "grassroots" organizations; that is, mostly small-scale, usually neighborhood-based activities, some of which are linked to the local government (volunteer organizations that stage organized sports on public playing fields) and some that are not (tenant associations in apartment buildings). In any city, much occurs politically that is not directly connected to or only tenuously linked to the decisions made and policies passed by elected officials. These activities flow from fraternal and religious organizations, business and neighborhood associations, labor unions, and advocacy groups involved with such local issues as historic preservation and living wage laws and national issues as immigrant rights and abortion. The overarching purpose of these governance arrangements is to bring democracy to everyday life.[11]

From the start, when villages and towns became big enough to be thought of as having neighborhoods—that is, as having residential areas distinct from each other—their inhabitants formed organizations that protected them from harm and reinforced their sense of belonging to these particular places. Probably the earliest of these organizations were the churches around which settlers and, later, immigrants clustered. These were not solely places of worship. Rather, their ministers, priests, and later rabbis and imams frequently became involved with births, marriages, and deaths; passed judgment on the morality of public and private actions; counseled newcomers; provided food and clothing to those in need; and joined the neighborhood to other parts of the city and to the local government.

Religions gave rise to benefit societies like the Knights of Columbus in the Catholic Church and B'nai B'rith, a Jewish community service organization. Similar organizations formed outside of a religious context (for example, the Freemasons and the National Association of Colored Women's Clubs) and became involved in community work by providing free lunches to the homeless or helping single women to find places to live. They sponsor sports leagues and hold charitable events. In many of them, people are elected to office and members have a say in how they are run.

Less spiritual but no less important for urban governance are labor unions. Throughout their history, they have taken on community as well as workplace concerns. In Milwaukee in the early twentieth century, labor unions were strong and deeply engaged with life outside the factory. Prior to the 1950s, Milwaukee was a working-class city with highly unionized machine shops, automotive supply plants, electrical equipment manufacturers, and breweries. The unions were politically active. They supported Socialists running for political office and organized immigrants to vote in municipal elections. When elected to office, Socialist and pro-labor politicians pursued strict regulations on private businesses, the provision of affordable housing, a debt-free local government, and an extensive network of parks and neighborhood social centers. They also championed such working-class leisure activities as bingo, gambling on games of chance, and pinball and resisted the efforts of middle-class reformers to ban or severely curtail them and the taverns in which they took place. Unions supported the desegregation of bowling leagues and generally favored racial egalitarianism. Civic events such as parades were also a major part of the union presence in Milwaukee. Labor unionism was social unionism.[12] In other cities, labor unions defended workers in their jobs, pressured management for higher wages and better working conditions, and engaged in the lives of their members beyond the confines of the workplace.

By the twenty-first century, the center of union activity had shifted from manufacturing to public services, union membership had declined significantly, and conservatives at the federal level and in state governments oppose labor organizing and suppress labor unions. Still, union activity continues. Construction unions provide political support for infrastructure and redevelopment projects, teacher unions engage in local school policy, and municipal unions put their resources behind wage legislation. In Baltimore, Los Angeles, Albuquerque, St. Louis, San Diego, San Francisco, and numerous other cities, local coalitions involving municipal labor unions have succeeded in passing ordinances that mandate the upgrading of low-wage jobs by paying employees an income that will sustain a decent standard of living, an income usually well above the federal minimum wage. This living wage initiative has functioned as a movement-building tool that brings together labor, community, and religious groups in ways that draw on and strengthen democracy.[13] Regrettably, unions have a dark side. Construction, police, and fire unions have historically resisted giving membership to women and minorities and labor unions are notorious for their oligarchic tendencies leading, in some instances, to corruption.

Their more beneficial activities are never as clear as when unions join with local communities to advocate for public benefits from large, ostensibly private construction projects.[14] Most often negotiated with developers, with local governments at times serving as intermediaries, the results of these efforts are known as community benefits agreements (CBAs). Such agreements are meant to ensure that the residents of the city most affected by redevelopment initiatives receive the benefits from them. The best known of these agreements was signed in Los Angeles in the early 2000s after unions, local residents, and housing groups organized to assure that a major commercial development near the downtown would have sufficient local benefits to overcome the costs (for example, increased congestion) imposed on those living nearby. The coalition negotiated with the developer for affordable housing, funding for parks, a "living wage" agreement, and local hiring. It intervened to do what the local government was not doing; to wit: assure that residents were compensated for the disruption caused by a major development initiative. Not to go unmentioned, such community benefits agreements can also be manipulated—the process of devising them undemocratic—and insensitive to larger public concerns.

Business and resident associations have also been part of the governance of cities for decades. Since the early twentieth century, and in the retail districts of neighborhoods and downtown areas, business owners have come together to lobby for public improvements and advertise to attract shoppers. Chambers of Commerce and Boards of Trade are examples. The Greater Miami Chamber of Commerce in 2015 had a staff of 25 professionals and numerous committees to address community and industry growth, government affairs, international business, and leadership. Moreover, it paid close attention to and involved itself in deliberations over downtown development.[15]

Voluntary business groups are also ubiquitous. Their goal is usually to preserve and enhance the viability of neighborhood retail districts. They attempt to speak with one voice to the local government, while watching over the district to assure that it is clean, crime-free, and attractive to shoppers and other patrons. This not only serves the businesses, but also stabilizes the neighborhood. As one example, the Old Town Boutique District in Alexandria (VA) consists of the owners of over 30 jewelry, food, clothing, wine, and home furnishing stores and works with the local Chamber of Commerce to keep the district a desirable place for residents and tourists. Across town, the Del Ray Business Association undertakes similar activities for the commercial strip of Mt. Vernon Avenue.[16]

A more recent manifestation of such place-based associations are business improvement districts or BIDs. These are organizations of commercial property owners located in a specific area of the city. A special tax on their property is used to fund improvements (such as signage and street furniture), provide additional maintenance of sidewalks and plazas, and even offer security services.[17] Essentially, a BID is a service-providing body sanctioned by the local government that augments the public services already being provided. BIDs require municipal legislation, are governed by boards, and coordinate their activities with local governmental agencies. The Bryant Park BID in New York City, to take one instance, oversees a major green space that is available for sitting, eating lunch, and socializing. The BID maintains this park, provides outdoor furniture, programs activities such as nighttime movies or concerts, and works with the surrounding businesses to keep the area clean and safe. To be clear, BIDs mainly serve property owners and enhance their properties. The improvements that are provided, though, do serve a larger public. Yet, they function with little resident input. Consequently, not only is public accountability an issue, but their undemocratic nature as well. Still, and like Chambers of Commerce and other associations, they are an integral part of urban governance.

Harboring the potential for democracy and oligarchy are the homeowner or resident associations that exist in neighborhood after neighborhood across the cities of the country. These entities often form when a neighborhood is threatened or falling apart. They are a product of the mid-twentieth century and came into existence as residents concerned that their neighborhood might be severed by a highway, burdened with a scrap yard, or invaded by people unlike them organized to protect their homes and businesses. The prototypical examples are homeowner associations that resist neighborhood change whether that change takes the form of blight and a decline in property values, a changing racial make-up, or a government initiative that imposes an unwanted land use (e.g., a waste transfer station) on their neighborhood.

Homeowner associations are also common in cities with apartment buildings comprising owner-occupied units. There, the residents purchase their apartments, join in either condominium or cooperative arrangements, and engage with management companies to oversee the operation of the building, decide questions of maintenance and repair and renovation, set common fees, and, for cooperatives, screen potential buyers and thus those who will become their neighbors. Counterparts can be found in rental apartment buildings where tenant associations collaborate with management to assure that the appropriate services are in place and the building is maintained, and in public housing where tenant associations work with the local public housing authority to assure a safe and healthy environment. These associations (along with their buildings) are usually termed common-interest developments (CIDs), a term that applies to all entities that regulate (or govern) the use of private residential and related public spaces.[18]

CIDs proliferated in the late twentieth century with the rise of suburban, gated communities; that is, in popular terms, relatively large and affluent communities surrounded by walls or fencing and with gates (sometimes with guards) to control who can enter. Many of the gated communities, however, are actually occupied by middle-income and low-income renters and lack elaborate security provisions.[19] Regardless, owners and renters band together to govern the use of common spaces, maintain individual and shared spaces, and even regulate social behavior (for example, whether front lawns can be used for social events). The more affluent and larger gated communities also provide trash collection, street lighting and maintenance, and security—services that are usually the responsibility of the local government. In addition to paying local property taxes, homeowners and renters pay a yearly fee for these privatized services.

The general intent is to maintain the character of the community or apartment building, bolster the value of the property, and ensure safety. In many ways, this is a contribution to local democracy. People are organized into governing bodies that develop and oversee regulations and make decisions regarding the upkeep of common spaces. The emphasis on property values, though, can shift the focus away from "living together" and that on safety can make the community fearful. As an anomalous example, in 2012 a member of a gated community crime-watch shot and killed a young, black man (Trayvon Martin) who was walking through the complex. The resultant furor and acquittal of the shooter led to a social movement, Black Lives Matter, whose purpose is to publicize and resist anti-black racism pervasive in the United States.[20]

A quite significant contribution to neighborhood governance has been made by the neighborhood associations that arose in the aftermath of World War II when many city governments embarked on highway programs to relieve traffic congestion, demolish blighted properties and slums, and build sports stadiums and convention centers. Slum clearance directly affected low-income (often African-American) residents while highways, redevelopment projects, and new public facilities (such as convention centers) were frequently sited adjacent to or overlapped with industrial districts and inner-city neighborhoods. These initiatives were almost always detrimental for the social relations and homes of residents and businesses there.

Although such initiatives met little resistance in the early 1950s, a few years later, residents and business owners organized to oppose them.[21] Marching in protest in front of city hall, linking arms to block bulldozers from demolishing buildings, testifying at public hearings, publicizing their concerns in newspapers, and cajoling elected officials, they acted to protect their neighborhoods from being sacrificed to what planners and the government, but not they, considered progress. Redevelopment and highway initiatives displaced businesses and forced residents to find homes either in more expensive areas or in areas that were becoming overcrowded. Most residents lived where they did because they wanted to be close to people—Italians, Puerto Ricans—of similar nationality and ethnicity. There, they were comfortable with stores, restaurants, and services that fit their needs. Even African-Americans and Puerto Ricans who faced discrimination in the larger housing market and were forced to live in less desirable neighborhoods did not want to cede what was positive about living with each other. Their residents had social relations and memories that were of value to them. Neighborhood associations acted to protect these attachments and their supporters viewed such opposition as morally defensible even if practically infeasible, given the elite politics surrounding renewal efforts.

Probably the best known of these struggles are those described in Jane Jacobs' famous *The Death and Life of Great American Cities*. In it, she defends fine-grained and diverse neighborhoods like the West Village in New York City, where she lived in the 1950s and 1960s. Jacobs encouraged residents of this and other neighborhoods to organize in opposition to governmental incursions that would displace existing residents in order to build middle-income apartment houses, widen streets to accommodate automobiles, and bisect local parks with limited-access highways.[22] In New Haven, Mayor Richard Lee in the 1950s spoke of his dream of a "slumless" city and turned to the federal government for financial and legal support to achieve that goal. One of his targets was the African-American Dixwell Avenue neighborhood, where the City bought blighted properties, evicted tenants, and demolished buildings. In response, residents organized and attended city council meetings to voice their opposition, publicly protested, and joined with the National Association for the Advancement of Colored People (NAACP) and the Congress of Racial Equality (CORE) to pressure elected officials to take a less disruptive path to mitigating slums and improving the neighborhood. Large-scale clearance continued, even as the municipal government built a new library, schools, and housing and engaged in neighborhood beautification. Most of the new housing, however, was beyond the financial means of existing residents and, while many residents left voluntarily, many others were forced to leave.[23]

Neighborhood resistance to disruptive governmental initiatives was strengthened when then-President Lyndon Johnson launched his War on Poverty in the 1960s.[24] Federal legislation enabled poor neighborhoods to set up Community Action Agencies to address a deficit of educational, policing, health, and recreational services and to advocate before local governments for greater attention to their needs. These agencies organized local residents, engaged in neighborhood planning, and even provided a platform for political activities. With the cessation of direct federal financing in the 1980s, many of these agencies endured in another form. Some received funding from the local government to continue operations. In Philadelphia, the City distributed small grants to volunteer associations so that they could monitor their neighborhoods and work with the city to provide services and upgrade their surroundings. Others became nonprofit Community Development Corporations (CDCs) and took on the construction of affordable housing, operating such businesses as supermarkets, and providing day care and job counseling. Another federal program, Neighborhood Housing Services (NHS), created a similar form of community-based governance.[25]

These placed-based organizations are found throughout cities in the United States, particularly in low-income neighborhoods. One of the most famous is the Dudley Street Neighborhood Initiative (DSNI) in the Roxbury area of Boston.[26] DSNI was established by residents in 1984 when almost one-third of the land area was either vacant or occupied by abandoned buildings, arson was common, and residents suffered from high levels of poverty and unemployment. The intent was to take control through bottom-up participatory planning coupled with such activities as organizing sidewalk improvements, providing day care for the children of working families, constructing affordable housing, and mounting voter participation drives to increase their political influence at City Hall. By 2015, DSNI had built over 400 housing units, renovated nearly 500 homes, and rehabilitated over 1,300 vacant lots while growing to 4,000 members.

Less deserving of praise are those neighborhood and homeowner associations that have blocked new groups—people unlike their members—from taking up residence. In almost all cases, these have been white, working-class neighborhoods whose residents have felt threatened by the encroachment of African-Americans.[27]

In Detroit during the 1950s and 1960s, white homeowners resisted the influx of black families; an activity one scholar termed *defensive localism*. Approximately 200 such associations formed with various labels—civic, protective, improvement, homeowners—to assure that neighborhoods would remain white. Large numbers of African-Americans were migrating from the South to the Northern cities and significant racial change was underway. Homeowners were fearful that their way of life would be trampled and their property values diminished. They monitored the sale of homes to black families, pressured local elected officials to withdraw support from civil rights organizations and support restrictive covenants that limited sales to white families, and attempted to intimidate real estate agents who were engaged in these sales. When these efforts failed, they picketed the properties of black families, threw rocks and garbage at the newly occupied homes, and destroyed fences and lawns. In a number of instances, their activities escalated to arson and physical harm. Moreover, it was rare for the police to protect African-American households. And, while many black families were discouraged from even contemplating a move into a white neighborhood, others made the effort. Many eventually "moved out of defended neighborhoods because of concern for their families' safety [while] others remained in their houses despite the harassment, and waited for the violence to abate."[28]

These associations have not been confined to low-income and minority neighborhoods opposing physical and social disruption or working-class neighborhoods resisting racial integration. Nor are they only organized to resist threats. They also exist in many affluent neighborhoods where their members beautify the neighborhood by planting flowers and trees, organize events at playgrounds, lobby for historic designation, and work with municipal agencies to improve services—all done to maintain the neighborhood's character and property values. In Philadelphia, the affluent Society Hill neighborhood is watched over by the Society Hill Civic Association that works with developers to ensure that new construction on vacant sites and the rehabilitation of existing buildings fit with the neighborhood's architectural style. It also advises the city government on requests for zoning variances and addresses parking issues. In upscale Nob Hill in San Francisco, the Nob Hill Association and the Nob Hill Foundation are dedicated to preserving and improving the neighborhood. If threats arise, these neighborhood associations—already well-organized and well-funded—mobilize to resist them.

Neighborhood governance extends well beyond these various organizations and associations. It also occurs spontaneously and informally. This is apparent in shrinking cities like Detroit where the city government—fiscally strapped—is withdrawing public services from neighborhoods.[29] Absent municipal street-cleaning, residents remove the trash themselves. They board up abandoned buildings and pick up the junk mail scattered on the porches and lawns of empty houses. They set up crime-watch groups, encourage friends and acquaintances to buy homes in the neighborhood, and clean up the local playground. In short, residents take responsibility for the civic realm. In doing so, they contribute to a collective understanding of their neighborhood and establish a basis on which to build democratic practices.

Cities are also replete with organizations that advocate on behalf of one or another public issue (for example, the humane treatment of animals or the preservation of an historic site) or serve a specific group such as at-risk youths or military veterans. They include organizations fighting for a local living wage law, coalitions offering counseling services to victims of AIDS, groups opposing housing discrimination and police brutality, and neighborhood associations hoping to stop the turning of working-class neighborhoods into middle-class enclaves. These organizations often focus on issues that affect multiple neighborhoods, not just the ones in which they are located. Some have concerns that are national or international in scope. In Denver, typical of the country's cities, are hundreds of such organizations including La Raza Youth Leadership, Bridgeway (a home for pregnant teens), Bessie's Hope (for older adults), the Women's Global Empowerment Fund, the Denver Art Society, and the Greenway Foundation concerned with environmental preservation.[30] Big cities in the United States are likely to have local chapters of the National Association for the Advancement of Colored People (NAACP), Amnesty International, and the Gray Panthers. (See table 4.1.) Their residents benefit from having these associations and organizations in their midst, not only as regards the services that they provide and the protections they hold forth, but also the many ways they involve residents in democratic practices and instill a sense of belonging, a sense of being together in the world.

Broadening our gaze reveals cities as places where social movements can take root and thrive. The Temperance Movement of the early twentieth century focused its efforts to ban alcohol consumption on the many saloons and taverns that existed in immigrant, urban neighborhoods. The second wave of feminism benefited from rallies held in the city's public spaces, consciousness-raising groups that met in apartments, and such "safe" spaces as women's health clinics, bookstores,

Table 4.1 Selected Advocacy Groups in Seattle

American Civil Liberties Union
American Friends Service Committee
Amnesty International
Audobon Society
Committee Against Rape and Abuse
Earth Save
Food Not Bombs
Gray Panthers
Jobs and Justice
Lifelong AIDS Alliance
Northwest Environmental Watch
Planned Parenthood
Sierra Club
Statewide Poverty Action Network
Washington Physicians for Social Responsibility

Source: Seattle Activism website, accessed February 15, 2017, www.seattleactivism.org/links.asp.

and community centers. The civil rights movement of the 1950s and 1960s involved marches in Selma, Alabama, and huge rallies in Washington, DC. Rosa Parks, a civil rights activist, refused to cede her seat to a white man in Montgomery on December 1, 1955, and the subsequent bus boycott and court cases resulted in the US Supreme Court declaring such segregation unconstitutional. The movement to protest social and economic inequality in 2011—Occupy Wall Street—began in Zuccotti Park in New York City and spread to Boston, Lincoln (NE), Seattle, Milwaukee, Chicago, and Salt Lake City. Black Lives Matter began in Oakland.

Cities offer a fertile environment for grassroots movements to flourish.[31] They contained these and numerous voluntary associations, some focused on members (for example, rod and gun clubs, veterans' organizations, ethnic lending organizations, softball leagues) and others oriented to a larger community as with gay health clinics, shelters for victims of domestic abuse, and youth drop-in centers. Together, these various organizations and associations represent an important layer of governance in which people gather to give meaning to their lives and, in the process, govern communities.[32]

Democratic practices are everywhere in cities. Democracy thrives where numerous opportunities exist for people to help others, come together to address concerns about which they feel strongly, and engage in activities that enable them to live well together. Many of these efforts are connected in some way to the local government (as with governmental funding that flows to neighborhood associations), but are beyond the direct control of elected officials and governmental bureaucrats. Others are wholly independent, like a bowling league or a local historical society. A few (for example, anti-immigrant groups) are driven by fear and intolerance.

Local Government

Only the most informal of these various associations escape the shadow of government. Nonprofit organizations have to abide by tax regulations that limit their political activities and stipulate how they can use their revenues. Book groups convene in the public library and soccer leagues use municipal playing fields. Neighborhood associations and business improvement districts receive public funds to support their services. Many organizations, as with volunteer associations that provide counseling to refugees, work with governmental agencies such as housing authorities and health clinics to meet the needs of their clients. These organizations are deeply enmeshed in civil society, but the boundary between civil society and government is blurred. That between the government and the economy is no less fuzzy and along this frontier is where we discover the relations that nurture the city's oligarchic tendencies. To this extent, much of what needs to be said about the oligarchic and democratic tendencies of local government revolves around its relationship with businesses and its commitment to economic growth.

Few observers would deny that US cities create the conditions for concentrating power and political influence—not just wealth. Confronted with various demands to ensure economic growth, protect the environment, attract new residents, and deliver a range of public services to residents and even tourists, local governments are organized to attract economic investment and, as part of doing so, solidify political control and constrict access to governmental decision-making. Yet, local governments and the officials who run them cannot summarily dismiss legal requirements to hold open elections and allow public scrutiny or abandon claims to legitimacy. A democratic impulse exists, but in most cities it struggles to maintain itself in the face of a tendency to narrow political access, serve the interests of the business sector, and respond to organized constituencies.[33] Local governments command resources—not just funding but also laws, regulations, and jobs—and have the capacity to generate numerous business opportunities, hire workers to manage computerized databases and trash removal services, and respond favorably to the demands of various groups. These conditions work against the grain of democratic practices and enable people to pursue power, prestige, and profits.

Of course, elected officials cannot disregard the need to be reelected, or elected in the first place. Neither can they ignore the need to make public services responsive to their constituencies in order to be seen as legitimate. To serve these ends, and since the late nineteenth century, political parties have established a presence in the many communities that compose the city, usually doing so at the neighborhood level. These activities reached their full development with the rise of political machines, one of whose purposes was to gather enough votes that the machine could win elections. In order to assure this, the machine established neighborhood-based or precinct-based organizations, often with clubhouses, that not only connected residents with jobs (most frequently in government) but also provided them with food on holidays, among other favors. The precinct captain made it his job to ensure that whatever benefits the local government had available were dispersed to those who voted for the machine's candidates.[34] In addition, political machines often protected immigrants and Catholics from Protestant nativists whose intolerance often denied the newly arrived residents jobs, decent housing, and legal protections. Hardly democratically run organizations, they also fostered political engagement (albeit limited) and offered numerous supports for managing life in the city's neighborhoods.[35]

Political machines are much less prevalent in the twenty-first century than they were 100 or so years ago, although a number of them (for example, in Chicago, Albany, Philadelphia) survived into the post–World War II era. In Chicago, Richard J. Daley effectively controlled the local government from the mid-1950s to the mid-1970s through the use of his mayoral powers and his position as head of the Cook County political machine. He was able to mobilize voters, centralize power, and reward supporters to govern effectively at the margins of democracy. Those mayors who replaced him, with the exception of Harold Washington (1983–1987) who attempted to "open" government to its citizens, were no less dedicated the needs of economic elites. Political machines, in the terminology of political scientists, evolved into growth machines.[36] Tellingly, in 1989, Richard M. Daley, the former mayor's son, was elected mayor and he ruled the city as tightly as his father but with less of the taint of political corruption and (white) ethnic favoritism. He served five terms until 2011.

That mayors in a number of cities have stayed in office for long periods of time is a testament to their ability to mobilize their supporters, satisfy powerful groups, and stifle opposition. Their extended tenures also increase the potential for democratic practices to be muffled. As one example, Mayor Michael Bloomberg (2002–2014), during his second term in office, managed to change the city's charter to allow him to run for a third term, which he won. In Boston, Thomas Menino served from 1993 to 2014 and in Providence (RI) Vincent Cianci Jr. held the mayor's office from 1975–1984 and then, again, from 1991–2002.[37] Other cities have had more frequent turnover with Albuquerque residents electing nine different mayors between 1974 and 2015. Long tenures do not automatically lead to the strengthening of oligarchic practices, and might even generate stability and allow long-term projects (for example, a park system or an overhaul of the police department) to be carried out, but they also have the potential to do otherwise, and often do.

Even if long mayoral tenures do not inevitably foster graft and corruption, they seem as if they would do so simply to maintain a steady stream of political support to remain in office. Such tenure is also likely to breed a sense of privilege and invulnerability. Mayor Richard J. Daley was constantly accused of illegalities occurring within his administration and the county political machine that he headed. The former mayor of Harrisburg, who served in office for 28 years, was indicted in 2015 on charges of diverting public funds to purchase museum artifacts for his personal use. Other mayors and elected officials have been charged and convicted even when they have served for only one or two terms. Camden, New Jersey, witnessed three mayors sent to jail between 1981 and 2000. And, it is not just elected officials who behave inappropriately. One hears frequently of public employees who take city vehicles for private use, steal from their departments, and reach into public funds for personal purchases. Agencies such as the traffic enforcement and the buildings department are particularly prone to corruption. Of course, we lack the evidence to know how prevalent corruption is and we cannot say that big cities are more prone to it than small towns. Corruption poisons democracy by making citizens cynical, and undermines the transparency that makes democracy work.[38]

City governments also have numerous mechanisms to make the political process, and the government, more democratic, particularly when they decentralize service delivery to the neighborhood level. Fire departments establish stations throughout the city and police departments divide the city into precincts, each with its own precinct house. Elementary schools are almost always neighborhood-based and recreational facilities (for example, playgrounds, swimming pools) are similarly dispersed. By doing so, the managers of these services expose themselves to the concerns of the area's residents. Police precincts have community liaison officers and elementary schools have parent-teacher associations that raise funds for extracurricular activities. A number of cities have even established official bodies to coordinate neighborhood service delivery, thereby matching, for example, after-school programs for teens with the availability of space in recreational centers. New York City in 1975 divided the city into 59 community districts and funded volunteer community boards (with small staffs) in each of them.[39] These community boards review zoning applications and watch over service delivery to ensure that the concerns of residents are being addressed. In Nashville, the city uses its 35 council districts as service districts. Looking more broadly, cities have numerous boards and commissions on which citizens can sit. Austin, Texas, in 2015 listed 73 such entities including the Historic Landmarks Commission, the Library Commission, the Human Rights Task Force, the Design Commission, and the Airport Advisory Commission.[40]

Public services, however, are not—in a straight-forward technocratic fashion—simply delivered in response to citizen needs and demands. These are political decisions mediated by organized interests, civic associations, and municipal unions. Take transportation as an example. Transportation Alternatives in New York City, a voluntary organization, has been quite influential in working with the city government to provide bike lanes, strengthen traffic laws that protect bikers, and provide accommodations for bike parking. It has a greater voice in these matters than the average citizen. The Olmsted Parks Conservancy in Buffalo (NY) has more of an effect on park policy than adjacent homeowners could separately. Police, fire, sanitation, and teachers' unions can influence the allocation of public funds for new facilities and decisions about how jobs are organized, thus affecting the quality of service delivery. Bus drivers can engage in work slow-downs or threaten to strike, thereby bringing pressure on elected officials to meet their demands but also disrupting the lives of bus riders. They can also advocate for more frequent service. Construction unions are a strong lobby when it comes to infrastructure investment such as bridges and water tunnels and large development projects. The actions of these unions serve the interests of their members and, at the same time, often improve services for all users, for example when teachers' unions call for smaller class sizes and more funding for libraries. The point is not that unions and organized interests are undemocratic elements of governance, but rather that they are part of the mix of oligarchic and democratic tendencies.

Where cities do tend to become oligarchic is in relationship to government contracts and the public incentives used to encourage capital investment, whether the relocation of a professional sports team to the city or the development of a former waterfront industrial site for apartment buildings and retail stores. This shift from a managerial to an entrepreneurial style of governance begins with the resources that local governments have at their disposal.[41] City governments are significant organizations with large budgets, big labor forces, and many different services from dog licensing to express buses and water purification. They also have innumerable laws and regulations that have to be monitored and enforced and that require information-gathering technologies and legal venues such as courts to adjudicate them. In addition, city governments are often deeply involved in fostering economic development and constructing such infrastructure as water tunnels and international airports. Every year they advertise for and give out large numbers of contracts for school supplies, gravel, printers, asphalt, office space, and police vans among the many items that a city government needs to function.[42]

City governments have an abundance of procurement opportunities to distribute. While the granting of contracts has to conform to legal limitations designed to prevent their being given to friends and acquaintances for political reasons, the size of many of these contracts favors certain businesses and not others. And, once a business has performed well and established relations with an agency, it is likely to be well positioned to have the contract extended and to be granted additional contracts. Graft and corruption are not the issue here, but rather performance and comfort with firms that have met the contractual requirements. This makes it difficult for new firms to successfully bid for business. Local governments recognize this and many have developed guidelines that compensate small and minority and women-owned firms for any disadvantages they might face. Numerous cities have set-aside programs to assure that these firms are able to compete, often allocating a percentage of contracts to them or using a scoring system to minimize bias against small and minority businesses. These efforts are minor compared with the overall tendency to maintain good relationships with large businesses, particularly when it comes to privatizing such services as trash collection. It is common to hear of politically connected businesses landing big contracts. Whether political connections were the reason is not always clear.[43]

While large contracts for supplies or services are one part of the concentration of influence, even more central to governance are the many opportunities that cities create, and local governments manage, for large-scale infrastructural and development projects.[44] Through its zoning and land use regulations, city governments create opportunities for constructing office buildings, apartment towers, retail malls, and factories. Real estate developers depend on local governments to modify zoning regulations to allow for greater density and thereby make existing buildings candidates for replacement and newer and larger buildings financially feasible. Or, the government might decide that a derelict or underutilized part of town should be developed with a large, mixed-use project and, subsequently, solicit bids from developers. Once again, the city government is generating opportunities and, most importantly, almost always subsidizing them by reducing the taxes that developers have to pay, providing construction loans at less than the prevailing interest rate charged by commercial banks, extending city services (for example, transit connections or sewers) to the site, and even assembling the site itself through the use of eminent domain. Just by extending sewer and water services, light rail lines, or bus service, city governments create investment opportunities. And when the local government decides to build an international airport, a new subway line, or a civic center or mount a climate change initiative focused on building infrastructure, the benefits for businesses are significant.

Not all firms can take advantage of these opportunities. The big developers, banks, insurance companies, construction firms, real estate brokers, and property owners all stand to profit. The small ones and those without political access do not. Elected officials are committed to growth and, for them, this means new construction whether it is office buildings or waterfront promenades. Since large projects are what garner the most attention and ostensibly attract other investors, thereby enhancing future prospects for growth, the larger the project the better. To the extent that the local government is a source of business profits, it is unsurprising that representatives of these firms are often politically connected, donate to the campaign funds of politicians running for office, and participate in various cultural activities (for example, buying tickets for a benefit dinner or sitting on the board of a museum) to signal their civic engagement. Public presence and personal connections are very much a part of this dynamic.

This dominance of large-scale public investment by a small group of property owners, developers, and investors is obvious as regards convention centers, a key component of the tourist-oriented development strategies of most US cities. These centers inevitably operate at a deficit with the local and/or state governments paying for the bonds used to build them and providing public funds to close the gap between operating costs and operating revenues. Those involved in their construction—construction unions, real estate lawyers, trash haulers, material suppliers—benefit. Nearby hotels can prosper. Those who own the property on which the center was built or who own adjacent property also do well. Residents pay. Moreover, the decision to construct or expand such centers is usually controlled by a small group of downtown property owners and investors. In St. Louis, decisions regarding where to locate a new convention center, whether to expand an existing one, and where to build a new sports stadium were all filtered through Civic Progress, a powerful business association whose goal was to protect and enhance their investments. Civic Progress was critical for organizing public support, obtaining state governmental funding, and assembling private financing. With large amounts of public funds at stake, it was not the voters who decided. This is typical.[45]

Control over development becomes even more concentrated when a city is dependent on a single industry or when groups coalesce to push a specific vision for what will make the city prosperous. Las Vegas is a good example of the former.[46] The city's economy is dominated by casino gaming and casino corporations, hotel owners, restaurateurs, airlines, and other service-providers whose livelihood depends on a robust tourist sector. Since the industry is highly regulated and the tourist industry itself requires public infrastructure from well-paved roads to large airports, those in the industry are politically involved in protecting and enhancing their investments. They have lobbied for improvements to the city's airport, the expansion of the convention center, and the upgrading of Fremont Street where the casinos were first established in the 1950s.

San Diego is another city where elite groups and institutions have come together around a shared agenda. There, the Chamber of Commerce; the University of California, San Diego; major businesses such as General Dynamics/Convair; and the Scripps Clinic and Research Foundation, among others, have combined forces to build on the presence of military installations to become a center for defense-related research and development. Joining them was the City's Economic Development Corporation. Together, they have established a cluster of science and technology firms that has added a third pillar to the region's economy of military installations and tourism. The city government, one of the many participants in this endeavor, provided land, extended business subsidies, and facilitated major redevelopment projects. The local economy was transformed in ways that favored educated labor. It was not done democratically.[47]

Politicians are well aware of the economic development function of local government, particularly those who want to make a career of public office. The path to influence is not solely one of proposing and supporting legislation that reflects the values of various constituencies. It is more fine-grained than that; it entails being aware of the many services and opportunities that government can deliver. Elected officials can decide to construct new high schools or not and do so in certain neighborhoods and not others. They can vote in favor of subsidies for a new baseball stadium or not. They can work with advocates for a living wage to pass supportive legislation or they can listen to business lobbyists and vote against the legislation on the grounds that it will make local businesses uncompetitive and cause them to leave. Depending on what positions they take, they can build a constituency that will ensure their reelection. Moreover, if it is done strategically, they can gain in influence by mediating between opportunities and those who want them. In many big cities, elected officials have authorized bond issues to construct stadiums for professional baseball or football teams in deals that directed all of the benefits to the team's owners and all the costs to the city's residents.[48] In other cities, mayors have made different decisions. The Mayor of Boston in 2015 withdrew his support from a private business group that was developing a bid for the summer Olympics. He did so in part because of the financial liabilities that would likely fall on the City government and its taxpayers, but he was also responding to opposition on the part of residents.[49] Once in office, elected officials can concentrate influence and limit democracy, or expand it.

One of the forces that constricts democracy is the ongoing support for initiatives that either privatize public services or turn the city's management of public services over to experts.[50] Privatization is based on the premise that governments lack the discipline of price signals generated by markets and are burdened by bureaucracy and interest-group politics. Consequently, they are unable to deliver services efficiently and effectively. Trash collection, animal control, water provision, and even public education would thus benefit from being managed and delivered by for-profit firms, or at least nonprofit firms that have to earn more than they spend. Public services would be isolated from politics. While the public sector would retain an oversight function, decisions about how to use labor and technology and who would be served would be left to the firms themselves, not to the government. Charter schools are a good example. These for-profit and nonprofit entities are meant to compete with public schools and with each other for students by providing the best educational services at the best price. Trash collection is another service that is often privatized, with public sector workers no longer doing this job and the city government no longer directly responsible for buying trucks and operating landfills. By itself, privatization is not necessarily oligarchic, removing what "must" be public from democratic deliberation. It might actually reduce costs for the government and improve performance. Nonetheless, it erects another barrier between citizens and those who provide essential services. Once privatized, these services frequently fall outside public deliberation and are accountable only to the pressure of consumer choice.

Privatization is a one-way street along which public services are transferred to the private sector, but the reverse also occurs. City governments can "municipalize" services previously provided by privately owned public utilities. Libraries in the United States were once private and now are mostly public. More contemporaneously, local governments can establish power plants as Sacramento has done with its electrical service. They can build city-owned telecommunication networks as did Alameda, California. They can also retain control over real estate that they own by leasing rather than selling it to developers, thus using it as a revenue-generating asset. At least one city (Harrisburg) owns a professional sports team—minor league baseball—and other cities have fan-owned teams: professional football for Green Bay and minor league baseball for Rochester and Syracuse in New York and Appleton in Wisconsin. Such initiatives counteract the trend toward privatization.[51]

City governments also deploy complex technologies that shift decision-making to experts. Computerized real property databases, crime reporting systems, public school student records, and zoning decisions are only a few of the various systems that are adopted. These technologies require expertise in civil engineering, curriculum development, public health, historic preservation, election and contract law, telecommunications, and plant biology. The current fascination with "big data" and "smart cities" is an extension of what has been a century-long imposition of expert systems onto local governments. The intent has been to use an array of data collection procedures such as sensing devices at bridges and tunnels, accounting practices, and the tracking of calls to complaint lines to monitor public service delivery and public needs in real time. The collected information can then be assessed by experts and decisions made as to where and when to allocate resources. If burglaries are increasing in one police precinct, police personnel are shifted there. If a storm has destroyed trees in one neighborhood, blocking streets, then crews need to be sent immediately to remove the debris. The underlying idea is that service delivery is a matter best handled by technocrats.[52] The complexity and arcane knowledge embedded in these technologies discourages widespread and active participation. Instead, residents become "consumers" who (in isolation from each other) express their concerns by calling complaint lines. Democracy is narrowed rather than broadened.

This does not mean that local governments ignore the need to be and be seen as democratic. As mentioned earlier, city governments, almost without exception, have numerous procedures in place to enable citizens to comment on and contribute to governmental decisions. Public hearings are held on a variety of issues, citizen commissions are appointed, city council meetings are open to the public, reporters "cover" local government and its many agencies, review boards are created, elected officials make public pronouncements, and legally binding regulations require that changes to zoning and land use regulations and infrastructure projects go through formal, citizen reviews. Elected officials have neighborhood offices to provide services to their constituents. During elections, current office-holders have to justify publicly their positions. Despite all of these efforts, and on the big decisions, access is limited to those with political or economic influence.

Intergovernmental Relations

To reflect on politics and governance only as they occur within a city's boundaries leaves unattended the multitude of ways that both play out across the political landscape. The governance of a city is more than the internal affairs of local government taxation and services, elite coalitions, neighborhood-based organizations, municipal unions, and fraternal groups. It also spans geographically across political jurisdictions and authorities encompassing not only the many municipalities that constitute the metropolitan area, but also relations between the local government, its state government, and the federal government. In some instances, it even extends globally when elected officials and local corporations forge ties with their counterparts in other countries to encourage trade relations or jointly develop city-based responses to climate change.[53] These forms of governance, as they drift further and further away from the daily lives of the city's residents and the places of the city itself, become less open to democratic engagement.

The history of intergovernmental relations in the United States is one where local governments were initially the most developed. Not until the twentieth century did the state and federal government take on significant responsibilities beyond, for the federal government, military defense and international relations. By the mid-twentieth century, intergovernmental relations had become a major component of local governance. States and the federal government have become more and more involved in regulating local taxes and services, providing funds for municipal services (such as policing), and mandating new services such as anti-terrorism task forces and lunch programs for children of low-income families. What did not evolve in a similar way was metropolitan governance that would mirror the functional linkages that tie cities to their suburbs.[54]

As the state and federal governments evolved, they met needs that had previously been ignored (for example, intercity highways, school funding, and retirement insurance). To provide such services they instituted new taxes on income, retail sales, and business profits.[55] The result is an overlapping of tax jurisdictions with services being funded and delivered from a number of sources. Residents are faced not just with a local government providing street maintenance, taxing them for this and other services, and creating venues for democratic practices, but multiple levels of government—municipal, state, federal, and others—offering an array of public benefits. Making the lines of accountability even more complex is the existence of governments (for example, counties) that sit between the locality and the state, school districts that do not always correspond with municipal boundaries, and regional taxing and service provision authorities involved with water and sewage, waste management, or mass transit. On a daily basis, the residents of the city might benefit from (and pay for) a road maintained by the state government, a bus service provided by a regional body, health care financed through a federal program, a library funded by the municipal government, and a police department drawing funds from all levels of government.

City governments are entangled in a web of intergovernmental relations. Of major importance are the intergovernmental transfers that flow to them from above. (See table 4.2.) Since the 1980s, state transfers to municipal governments have hovered around 35 percent of local revenues and federal transfers have been about 5 percent. For the

Table 4.2 **Intergovernmental Transfers for Selected Large Cities, 2006**

City	Percentage of Total Revenues
Boston	41.9
Philadelphia	33.7
Memphis	30.4
Albuquerque	28.0
Charlotte	21.8
Columbus	16.1
Nashville	13.8
Jacksonville	9.8
San Antonio	5.8
Los Angeles	4.7

Source: *Statistical Abstract of the United States: 2012.* Calculated from Table 457, accessed April 14, 2016, www.census.gov/library/publications/2011/compendia/statab/131.

average municipal government, then, approximately 40 percent of its revenues originate from outside the city. With these transfers come limitations on how these funds can be used; federal and state goals are meant to override local objectives. Of course, local governments can always refuse the funding. At times, higher levels of government require local governments to offer services such as screening children for a disease, but fail to provide funding to do so. These are known as unfunded mandates and further constrict the autonomy of local decision-making.[56]

That said, residents rely mainly on locally elected officials to engage with state and federal governments. Larger cities even establish offices in the state capital or in Washington, DC, where staff can monitor legislation, act as liaisons with relevant governmental agencies, and lobby legislators for various programs. City residents also have access to state and federal representatives, all of whom provide constituent services, and can vote to elect to or oust them from office. Of particular concern for any city government is the state legislature. City governments need to obtain its approval for any new taxes or for programs not granted to it in its charter. According to the famous court case written by Judge John Dillon in 1872, municipal corporations (cities) are subordinate to state governments; that is, they derive their powers from state legislation. Judge Dillon was concerned that city governments would stray beyond their public functions and take on private-sector activities. Consequently, if municipal governments wish to institute a new tax on automobiles entering the central business district to decrease congestion, a practice known as congestion pricing, or set up a public authority to undertake a large redevelopment project, they must receive state governmental approval. The problem many large cities face is that state legislatures are no longer dominated by city representatives as they were in the early twentieth century. Rather, suburban representatives are more numerous and often disinclined to favor city interests. The effect is to limit the options available to city governments for managing their fates.[57]

A similar type of relationship, absent the constitutional dependency, occurs with the federal government. Since the 1930s, a period of rapid federal government expansion triggered by a Depression and World War II, the federal government has launched a large number and wide variety of programs to address the concerns of the country's people and places. It has extended its purview beyond national defense, central banking, and trade policy to subsidizing businesses, funding policing technologies, passing environmental laws, and regulating mortgage markets. Many of these programs affect cities when they subsidize school construction or fund the highways that make sprawl more likely.[58]

What city governments hope to do by approaching Congress and federal agencies either directly or through their representatives is to influence how legislation is written and funding allocated. The government of New Orleans, for example, is quite interested in the plans of the US Army Corps of Engineers regarding the levees that channel the Mississippi River and affect the vulnerability of the city to flooding. Boston, in the 1990s, approached the federal government for funding that would allow it to take down an unsightly elevated highway (the Central Artery) and build a tunnel in its place, a project known as the Big Dig that eventually cost $14.6 billion. New York City in the 1990s used federal highway funding to demolish an outmoded and unsightly elevated highway and replace it with a tree-lined boulevard. San Francisco, Milwaukee, and Portland (OR) also removed limited-access highways from their waterfronts with federal assistance.[59]

Lurking behind these city-federal relations since the 1960s has been a belief that the federal government has a responsibility to cities that should manifest in a national urban policy.[60] The creation of the federal Department of Housing and Urban Development (HUD) in 1965 was a partial recognition and response to this belief. Other efforts have occurred to realize this ambition, but no policies have been passed that privilege cities as a special constituency in federal politics. In fact, the likelihood of a national settlement policy in the United States is miniscule given the restrictions on capital and residential mobility that would be required. Nevertheless, a number of states—New Jersey, Oregon, Maryland—have legislated statewide development policies.[61] Their primary intent is to utilize existing infrastructure, rather than build anew, and to preserve farmland and the undeveloped areas of the state. With the exception of Portland, Oregon's growth boundary, these efforts have had little success in encouraging more dense development within and close to existing urban centers. Doing so would ostensibly benefit the cities economically and politically and in this way enable them to prosper. A robust federal-level settlement policy would remove such decisions from local control. The emergence of stronger state governments and a federal government reaching down into municipalities has strengthened city governments by providing additional funding; it has also made them more dependent on other levels of government.

Less developed are relations among the many municipalities that compose metropolitan areas. Given the functional interdependencies that abound, a casual observer would imagine that elected officials would want to speak with and coordinate activities with their counterparts in nearby places. Coordination, in fact, is fairly common around regional issues such as mass transit, water supply, airports, and ports. Louisville has a Regional Airport Authority, New Orleans has a Regional Transit Authority, New York has a regional Port Authority, and Detroit has a regional Convention Facility Authority.[62] Special purpose districts, often crossing municipal boundaries, are another governance mechanism for achieving economies of scale in the delivery of public services. They are mainly used for water supply, sewage, flood control, and housing.[63] The Minneapolis-St. Paul region is known for its metropolitan tax-base sharing program meant to discourage intergovernmental competition for relocating businesses that drains revenues from local governments. A small number of metropolitan areas have consolidated city-county governments, some of which are metropolitan in scope: Nashville, Miami-Dade County, Indianapolis, and Honolulu among others. City governments in Philadelphia, Los Angeles, and San Francisco are coterminous with their county governments. And, New York City comprises five county governments, now treated as boroughs of the city. In general, however, city governments are rarely involved politically with their suburbs. This is due, in part, to a fiscal federalism that makes local governments heavily reliant on revenues generated within their jurisdictions by property and resident income taxes. Sharing is discouraged.

City governments look outward to the world as well. Almost always this is a matter of remaining economically competitive for trade, tourism, and dominance in one or another sector of the economy such as the visual arts, amateur sports, or financial services. It also entails the pursuit of foreign direct investment as when the local economic development agency offers financial and other incentives to a German, Chinese, or English firm to locate in the city. Elected officials are also enticed by international comparisons that rank their city in terms of economic competitiveness, creativity, livability, and global influence. For the larger cities, being "global" has a cachet that brands them as places of opportunity. To be a global city is to be formidable in the world of cities, but also to be compelled to vie incessantly for attention by investors, tourists, and highly educated migrants. Consequently, large cities advertise for tourists in international magazines, development entities send trade missions to other countries, and governments join various networks (such as the C40 Cities Climate Leadership Group) to share ideas and become globally visible. These activities frequently occur beyond the reach of democratic procedures and channels. They hardly ever build on the many connections that less educated and recent immigrants have with the country that they left. What matters is inward investment and finding trade partners for local businesses.[64]

Unlike cities in countries where the central government is primarily responsible for collecting taxes and redistributing it to localities, thus enabling a certain equity in revenue distribution, American cities are encouraged to compete against each other for affluent residents, thriving businesses, outside investors, and global prominence. This competition favors the inclusion in local governance of economic and cultural elites who have the connections and incentives to think and act beyond the city's boundaries. In this context, having a national and even global image is a resource. Being able to compete against other municipalities for state funding and programs and against other cities for federal assistance is viewed as desirable and spurred by the need to maintain a robust tax base. In external relations, participation by citizens is marginalized.

Conclusion

Living in close proximity creates consequences for others. Certain of these consequences are welcomed, as when a much-needed supermarket opens in a neighborhood or when street musicians decide to perform for an admiring audience. Others are less acceptable as when a nightclub keeps nearby residents awake or when the managers of an apartment building pile the sidewalk with trash. Procedures must be developed to manage these conflicts in ways that enable the city's residents to feel that their concerns matter and are being heeded. In response, democratic practices emerge at the grassroots level and local governments are formed to address citywide issues of road management, water provision, and public schooling. Newspapers, radio and television programs, and local magazines bring these issues to public attention, further encouraging democracy to prevail.

Within civil society, many of these activities are open to the involvement of numerous individuals and groups. Even then, a neighborhood-based association that does not stay in contact with and listen to the concerns of the residents will quickly lose legitimacy. Civil society can also be oligarchic. Local governments face a similar dilemma of how to balance the imperatives of large-scale organizational dynamics with the need to be accountable. But large, city governments—and all formal organizations—harbor a tendency to concentrate influence among a small group of individuals. The business opportunities that these governments have to dispense, the political opportunities for becoming influential, and the personal benefits of elective office make them susceptible to nondemocratic practices. The journey begins with the need for representatives and ends with elite coalitions that gather onto themselves the profits of government: zoning that creates real estate investment opportunities and contracts for building a new sewer line, to name just two such opportunities. The expertise ostensibly needed to manage technologies and the complexities of intergovernmental relations further reinforce this tendency. Oligarchy is in tension with the participatory mechanisms that local governments also support. Under pressure from citizen groups and the scrutiny of local media, governing elites attempt to balance the appearance and substance of democracy with interests often cast as beneficial for the city but which, when stripped of their celebratory rhetoric, serve only a few.

The story is not the same across all cities. Some are highly controlled by a small group of elites, while others (albeit too few) have nurtured progressive groups—strongly committed to democratic practices—that are active in both social and political life, even to the point of placing like-minded people on city councils and in the mayor's office. San Francisco is one such city where, in the 1980s, progressives were able to restrict downtown growth, elect gay activists to public office, develop policies to aid the homeless, and even elect a liberal mayor. Progressive movements have thrived (briefly) in Boston and Chicago, Santa Monica, Santa Cruz, and Burlington (VT). For those committed to democratic governments and just cities, maintaining positions of power has always been difficult. Even when successful at shaping city policy, they have had to contend with relentless pressure from commercial property owners, developers, bankers, local elites, and newspapers (the growth coalition) for public funds to be invested in downtown development and only later, once growth has been established, to attend to neighborhoods and less advantaged groups.[65]

When governance spreads beyond the city, it becomes more and more confined to representatives and experts who mediate between the local government and state and federal governments. That the elected officials of cities mostly ignore relations with municipalities within the metropolitan area speaks more to the laws under which they operate than to any rational understanding of the interdependencies in which these cities are embedded. Yet, urban governance requires that elected officials and various publics interact with each other and engage with entities outside the city, including agencies and legislators from supra-levels of government. Governance happens both within and beyond the city's boundaries. As it is extended geographically and organizationally, participatory democracy is increasingly displaced by elite-mediated and expert-based politics.

Commentators on urban governance are fond of quoting the famous German historical and legal scholar Max Weber. In writing about the city, Weber reminded his readers of the German expression "*stadt luft frei*" (city air makes men free).[66] He used the phrase to note how the formation of cities weakened feudal relationships by encouraging commerce beyond the city's walls and increasing anonymity, thereby enabling people to do other than work the land or act as servants to the king. My interest, however, is less in the city as a place where people can be free as individuals than as a place where people become more connected to and respectful of each other through various forms of collective governance. That cities have representative governments as part of this governance array does not preclude deeply democratic practices. And, nongovernmental democratic practices can thrive even when the local government falls under the control of local political and economic elites who treat it as a source of business profits and a tool for enhancing their political ambitions.

Cities enable both democracy and oligarchy. Seemingly opposites, these qualities exist simultaneously, interdependently, and inseparably. Elites take control of the major investment and policy decisions made by local governments, while citizens exert influence during elections and via a multitude of organizations that foster civic democracy and elevate participation above representation. Despite their efforts, the residents of the city cannot escape the consequences of decisions and actions over which they have little influence.

Given the political culture of the United States, democracy and oligarchy are not separate conditions occupying different spheres in the city's governance, but are intertwined. A city that is fully democratic is a dream, a utopia useful for imagining a better world, but still a utopia and unrealizable. It is similar with tolerance and intolerance, the topic of the next chapter. A tolerant city devoid of animosities, disrespect, and ill treatment is an ideal, possible only in our imagination. Nonetheless, it is a worthy goal to which to aspire. Like poverty and wealth, degradation and sustainability, and oligarchy and democracy, intolerance and tolerance are ever-present.

FIVE

Intolerant, Tolerant

Faced with isolation or even persecution, people often flee small towns and make their way to the big city. Once there, they hope to become anonymous or, at the least, escape unwanted scrutiny. Ideally, they would meet like-minded people with whom they might live and mutual support would follow. Being gay in a rural community where religious values condemn homosexuality or a Muslim refugee in a place populated almost wholly by nativist Christians can be psychologically uncomfortable and even dangerous. From this perspective, a big and socially diverse city looks to be a refuge.

Of the many reasons that people relocate to cities, one is the desire to be free from humiliation as they ride the bus, shop at the local supermarket, attend a concert in the park, walk through an unfamiliar neighborhood, and carry out their tasks at work. With cities, we associate tolerance. The expectation is that differences in religious beliefs, nationality, sexual orientation, ethnicity, political affiliation, race, lifestyle, and regional origin will be, for the most part, accepted or, at least, ignored. Those who are different will be treated according to criteria that apply equally to all. As the philosopher Michael Walzer has written, "Toleration make differences possible; difference makes toleration necessary."[1]

The differences that cities support bring the values and norms of diverse groups into contact. However, if that engagement is poorly negotiated, intolerance results.[2] In fact, acts of intolerance are ubiquitous. In almost any US city, we find areas segregated by race, ethnicity, and nationality. African-Americans, Mexicans, and Puerto Ricans in particular often encounter barriers to renting apartments or buying homes wherever they chose. In many neighborhoods, these groups are unwelcomed, regardless of whether they can afford to live there or not. Even worse, hate crimes against queers, African-Americans, and people from the Middle East are all too common, happening precisely because of where they come from and regardless of any objective threat that they pose.[3] Moreover, these are the very same people who experience discrimination by banks and insurance companies and in workplaces and retail stores. Not too long ago, women were denied home mortgages unless their husband or another male cosigned the loan and, in 2015, just after the US Supreme Court legalized same-sex marriages across the county, a debate ensued as to whether certain Christian store owners—opposed to such marriages on religious grounds—could deny services to same-sex couples.[4] At the same time, groups cohere by choice. In order to benefit from being with others like themselves, they establish a physical separation from other people. These arrangements constitute a form of voluntary social and geographical segregation that protects the group identities that are highly valued in a multicultural society like the United States. Not to be condemned, such segregation can also breed intolerance. Within cities, tolerance and intolerance coexist.

I think of tolerance and intolerance as political acts that signal a willingness, or not, to allow others into the public realm.[5] People different from each other come into contact and are faced with deciding how to behave toward those who are "other." Intolerant acts often have their roots in prejudice that, if deeply felt, slips into discrimination expressed in individual and collective (even institutionalized) laws and behavior that deprive others of rights and opportunities. A government intolerant toward Mexican or Syrian immigrants will define them as inferior, out-of-place, and without political rights, thereby denying them citizenship. By doing so, governments and the groups that support such policy preserve a specific and somewhat narrow understanding of the country's identity and engage in the constant imagining of the nation and its cities. Governments and groups tolerant of others contribute to more inclusive cities. And while tolerance has its limits, acknowledging those limits is integral to knowing where tolerance stops and intolerance begins. Politics is a matter of creating and delimiting publics and thus of indicating who belongs and who matters to the larger community. Tolerance and intolerance are two of its most powerful mechanisms.[6]

Although the city has "the capacity to teach its residents about tolerance," it also creates innumerable situations where intolerance, for some, seems to be the only option and—at times—a seeming necessity.[7] The city puts people with diverse religious beliefs and sexual orientations in close proximity, thereby forcing them into situations where they must compete for opportunities, confront (and often fear) encroachment by those unlike themselves, and judge each other harshly. Yet, the city functions best when conflict is minimized and people of various social standings, origins, and lifestyles are allowed to live unhampered by marginalization, humiliation, or violence. Tolerance and intolerance exist side-by-side. As a small example, consider that in many cities same-sex couples feel uncomfortable holding hands in public. Doing so makes their sexual preferences and lifestyle conspicuous and them, they believe, an object of verbal harassment or worse.[8] In many areas of San Francisco, Chicago, Los Angeles, and New York, this is not a problem. More "at ease on the streets," queer people there are mostly left to be themselves. Yet, they also know that to stray into certain neighborhoods—even in these cities—might put them at risk and a morally charged event (for example, a gay elementary school teacher accused of child molestation) could make public expressions of their sexual orientation ill-advised.

In this chapter, we consider how the city enables people—many of them strangers to each other—to live together both harmoniously and inharmoniously, asking them to negotiate the contradictory impulses of tolerance and intolerance. I begin with the basic idea that political acts of this kind are rooted in the social diversity of the city. Tolerance and intolerance are a response to "the other," whether to strangers, people of a different ethnicity or sexual orientation, or those with whom we simply disagree. After commenting on tolerance, the chapter then turns to different forms of intolerance with a focus on marginalization and violence. I end with a discussion of urbanity, a way of living that distinguishes life in the city from life in small towns and rural areas.

Diversity and Tolerance

Consider the many types of social engagements that exist within a city. In the early twentieth century, sociologists contrasted the city with villages and rural areas in terms of the mix of what they termed primary and secondary relationships.[9] Primary relationships are those that characterize families and close friends; they involve people of whom we have long histories and extensive knowledge. They are siblings or first-cousins or friends as far back as elementary school. Secondary relationships are those that occur when retail clerks serve customers who are unknown to them. The clerks have little knowledge about the customer and no emotional attachment. The families of these customers and how they passed their teenage years is a mystery. Nevertheless, this does not preclude a transaction as the customer purchases a loaf of bread or pays a fare upon boarding the bus. Nor does it stop people from giving directions to a confused tourist. By definition, a city, in contrast to a rural community, has a much higher percentage of secondary relationships. Another way to say this is that a city contains a greater proportion of strangers. In villages, few strangers appear; whereas in cities, the streets, shops, bus stations, and theaters are filled with them. Encountering someone we know in the city's central park, at a street festival, or in a downtown department store comes as a surprise.

Unlike a small village where many people are related by blood or marriage and, as long-term acquaintances, are familiar with each other, in cities many of the people one encounters are unfamiliar. If, as well, they are perceptively different in skin color, language, or dress —or considered radically different in ways that one or another group finds offensive—then tolerance is likely to be weakened. Because the signs of difference and the relationships that they evoke are malleable and often socially ascribed, tolerance and intolerance appear in divergent forms with differences sparking a variety of reactions. Certain differences—ethnicity, gender, nationality, sexual orientation, religion—are triggers for intolerance, despite numerous efforts to counteract these tendencies. Tolerance, then, requires sympathy for the diverse "other" and thus the "capacity to extend feelings, qualities, and visceral states" across divisions: it requires "artful experimentation with the social movements that draw individuals into togetherness."[10]

That said, cities are saturated with a variety of relationships: immediate family, relatives, friends, acquaintances, outsiders, neighbors, foreigners, colleagues, and strangers are all present. Such designations reflect how familiar we are with others, the obligations we might have to them (compare parents to business associates), and the amount of time we spend in their presence.[11] We might work most of the day with a colleague, but have no interaction with her outside of the office. We might know very little about a person, but in certain situations (for example, a parent struggling to move two children and a baby carriage onto an escalator) offer our assistance. The point is that these dissimilar relationships have implications for tolerance and intolerance. We are more likely to be tolerant of those with whom we have personal obligations, with whom we are familiar, and with whom we spend a good deal of our day. In primary relationships, tolerance is expected and intolerance an anomaly.

Of these various relations, two—strangers and neighbors—are central to understanding tolerance and intolerance in the city. To be a stranger is to be both anonymous and a curiosity and this status has positive and negative consequences. On the positive side, anonymity confers a certain protection from the judgment of others as the anonymous person is less likely to be pulled into a web of social norms embraced by people familiar to each other. Seemingly on the outside and beyond the prevailing social rules, the person is a stranger. Of course, this does not mean that these rules can be violated with impunity. At the same time, the stranger is a curiosity who might well engender trepidation. The actions and the consequences of an alien group detached from shared norms are unpredictable and, because of this, potentially disconcerting. What impacts will Arab newcomers have on the shops in the neighborhood, the schools, or how people use the local park?[12]

Despite such concerns, urban residents are to a great extent tolerant and even welcoming. Numerous cities have organizations whose purpose is to help strangers (often foreigners) adapt to the city. In Stamford (CT), Neighbors Link Stamford provides advice on schools, housing, and employment to immigrants; International Neighbors in Ann Arbor (MI) holds cultural exchanges and other social events to welcome women; and Littleton Immigrant Integration Initiative in Denver assists immigrants and refugees. In 2015, the mayor of Pittsburgh announced a set of initiatives meant to foster diversity and aid newcomers from foreign lands. His proposals included Welcoming Hubs at community centers and multilingual, employee-rights support. And, in Dayton, Ohio, the Welcome Dayton Plan pulls together job training, housing assistance, and business services and has established community centers in immigrant communities and improved public services to these groups.[13]

A useful and interesting distinction is between strangers and aliens.[14] Strangers are people whom we do not personally know but whose outward characteristics (for example, how they dress, their physical features) enable us to place them as belonging in the city or a neighborhood. In contrast, aliens are people whose outward characteristics perplex us—they are not from here. They are dressed in ways that seem outlandish, act "out of the ordinary," or are visually menacing. Aliens are less easy to read than strangers. It might be a group of Hare Krishnas or a leather-clad, tattoo-covered, heavily-pierced biker gang that shows up in downtown Boise, Idaho. Or, it might be Sikhs with their turbans and saris whose traditional garb makes them seem out of place. To be one of the first Somalian refugees in majority-white Omaha is to be an alien.

Seldom do we come into contact with strangers and aliens at home or at the synagogue. Rather, we encounter them in the public spaces of the city. It is on the sidewalks, in the parks, and at movie theaters and street festivals where we are likely to meet disparate people and observe behaviors that make us uncomfortable. It is almost only in public spaces that people physically meet the "other" whom they fear, resent, or condemn or who cause them to feel ill-at-ease. Of course, these others are also encountered on television and the internet, in magazines and newspapers, and at the movies, but from a safer distance. Tolerance and intolerance occur in public.

Diverse public spaces also have the potential to make people more accepting of others. Jane Addams, the Progressive Era reformer, believed that by encountering differences, people's sympathies were enlarged. Over time, diverse experiences make us more accepting. Public spaces—open to all for casual conversations, spontaneous group activities, and even protest—are central to democracy, itself a key aspect of tolerance.[15] Without places to deliberate together or to march for a cause, democracy is weakened and, with it, an openness to the opinions and presence of diverse others. Yet, it was in public spaces that civil rights marchers were heckled, spat upon, and even assaulted and, even more luridly, where nearly one-third of the mass shootings occurred in the United States in 2015.[16]

To simply divide the city into public and private spaces is to flatten the variety of spaces that constitute the city and in which tolerance and intolerance occur. Of major importance are parochial or semipublic spaces—airports, restaurants, shopping malls, and health clubs—in which the private and the public overlap and the distinction between the two is blurred.[17] University campuses are often this type of space. New York University, for example, is not isolated from "the city," in its own enclave, but intertwined with it. Campus buildings are interspersed with restaurants, apartment houses, pharmacies, and coffee shops. Through its spaces on a daily basis pass people who have nothing to do with the university. Students and nonstudents mingle on the sidewalks and in Washington Square Park, which sits at the symbolic center of this urban campus. The campus is a parochial space where a person is as likely to encounter a nonstudent as a student. It is a campus of many strangers and even aliens.

Most important for our discussion of tolerance and intolerance are the places we term neighborhoods. Neighbors are people who live together in an area of the city generally recognized as distinct. They tell people that they live in Over-the-Rhine (Cincinnati), the Garden District (New Orleans), or the Shepherd (Oklahoma City) neighborhood and the reference is obvious. Real estate agents, planners in the city's planning department, and dispatchers at the city's central post office all recognize these places. Because neighbors share a place, moreover, they also share an identity, whether it is personally adopted or conferred on them by outsiders. Other people in the neighborhood are familiar and look as if they belong, even if little more is known about them. A sense of community—a sense that neighbors should be acknowledged, respected, and given assistance if in need—also exists. Neighbors are "decent folk" and the inclination for most people is that they deserve to be acknowledged and given assistance if in need.[18]

As spatial identities coalesce, they also give rise to a feeling that the neighborhood has to be protected from unwanted change, whether involving an influx of dissimilar people or the imposition of a locally-unwanted land use such as a bus maintenance facility. In extreme cases, this defensive localism becomes violent as existing residents organize to discourage gentrifiers from moving in or minorities from buying or renting homes there.[19] Intolerance is the result.

This shared identity can also take a more tolerant form with neighbors coming together in communities of mutual support, helping someone simply because the person lives nearby. The elderly couple next door needs the snow shoveled from their sidewalk; the man across the hallway asks you to pick up any packages delivered to his door when he is on vacation. When Ida Mae Brandon Gladley moved from Mississippi to Chicago in 1938, her next-door neighbor came over to introduce herself. They shared a bottle of wine that the neighbor had brought and advice as to how to behave differently in the North than in the South.[20]

The status of neighbor is socially positioned between the alien and the family member. The attraction of neighborhoods for those concerned about cities is that it provides an intermediate realm between the public and the private such that people can find security and shared identity as well as have a presence among strangers. A relatively homogenous neighborhood can be a place of tolerance for the group and a source of social support. Neighbors are people who connect with those they only partially know and, when offering assistance and being perceived so by others, provide examples of neighborliness to be emulated. Neighbors can also be irritating and disruptive, encouraging avoidance or even public condemnation.

Of particular importance is the tendency of people to live in a neighborhood with people like themselves. Numerous neighborhoods are named for the group that has (or had once) congregated there: Greektown (Detroit), the Pilsen neighborhood in Chicago, Chinatown in Washington, DC, and Koreatown in Los Angeles are just a few of innumerable examples. Voluntary segregation provides distance from diverse others and contributes to a citywide tolerance even as it has the potential to exacerbate a more local intolerance. Yet, because few neighborhoods are never wholly homogenous, these residents are also constantly thrown together with people unlike themselves. They interact as patients at a local health clinic, customers at a neighborhood sandwich shop, strangers passing on the sidewalk, and participants at a rally against rising neighborhood crime, none of which necessarily forces them to acknowledge the ways in which they disagree or are different. The counter-person who serves you a sandwich at a diner might look Turkish and gay, but that has very little to do with enjoying your lunch and paying your bill. This is all quite unpredictable, though, and not readily generalizable. Whether the police officer who stopped you while you were driving because you were supposedly speeding is a Hindu likely is unimportant, but race might be important if the officer is white and you are a young, African-American male.[21]

Consider the diversity of just one neighborhood: Frogtown (from the German, Froschburg) in St. Paul. Today, it is known as a Hmong and Vietnamese neighborhood, but also living there are Somali, Mexicans, Chinese, and Laotians. Forty percent of the population is Asian with one-quarter African-American and another one-quarter white. Andersonville in Chicago was originally settled by Swedes and is still known for its Scandinavian origins, but is now a mixture of Swedish, Korean, Lebanese, and Mexican households and has a sizable queer contingent.[22] Implicit here is that ethnicity is a powerful basis on which groups cluster together.

One of the best-known of these neighborhood distinctions has its origins in the revolt against middle-class values and way of life centered on a striving for economic prosperity and social respectability. As embodied by writers, artists, intellectuals, political radicals, and feminists in the early twentieth century, this alternative to bourgeois ideals came to be known as bohemianism. In the United States, it led to the transformation of Greenwich Village in New York City from a proper middle-class community to a haven where bohemians gathered to live and socialize in artist studios, taverns, and the salons of wealthy patrons. Rebelling against what they viewed as a stultifying, middle-class culture, they nonetheless were symbiotically connected to it. Yet, living with the middle class would be to "sell out" and risk disapproval from their peers. In response, they sought out places with like-minded and like-acting people. After World War II, North Beach in San Francisco and Venice Beach in Los Angeles joined Greenwich Village as places hospitable to and tolerant of a bohemian lifestyle. In the 1960s, "beats" and "hippies" took up similar residence in places like Haight-Ashbury in San Francisco.

"For anyone black who sought admission, bohemia offered scant hospitality," however.[23] In New York City, black intellectuals, writers, artists, and political activists were found not in Greenwich Village but in Harlem. There, in the 1920s, black culture thrived to the point that the nightclubs where jazz was performed and places where poetry was read became destinations for white people—racial tourists—from downtown. This northern section of Manhattan became a community and safe haven for black men and women immersed in the bohemian life. Their presence created synergies that led to the Harlem Renaissance that crystallized a specific black urban culture of literary and artistic accomplishment and, for women, sexual freedom. Out of it emerged a racial consciousness that produced new forms of poetry, novels, and paintings that reimagined the black experience in America and elevated jazz to popular status in a certain segment of the white population. Black artists became "part of a group linked by ethnic pride, political activities, and a shared cultural lineage." The intolerance these individuals might have faced living elsewhere was absent.[24]

Beginning in the early twentieth century, bohemian neighborhoods also provided safety, security, and social support for queer individuals. Large cities freed gays and lesbians from mainstream constraints and offered the possibility of "surmounting loneliness." Queers created space for themselves in Park Slope in Brooklyn, the Castro district in San Francisco, Northampton (MA), Jamaica Plain in Boston, and Oak Lawn in Dallas among other places. Initially, they did so covertly but, later, made their presence publicly visible with bars, bookstores, and health clinics. If not tolerance, they wanted indifference; they wanted to live unbothered and outside the shadows. These places increasingly became known as queer spaces where holding hands or kissing in public was acceptable. Gays and lesbians also became politically active. They staged gay pride parades, established political organizations, and ran for office on platforms that championed issues directly related to the needs of the gay and lesbian community. Their numbers and presence gave these neighborhoods not just a public identity, but collective strength as well. Outside of them, the tolerance they enjoyed diminished. This has decreasingly been the case as the larger population has become more accepting. As just one example, and as mentioned earlier, in 2015 the Supreme Court legalized same-sex marriages across all states and localities.[25]

The contemporary version of the bohemian neighborhood is the "hipster" neighborhood. Many of the hipsters and artists attracted to these neighborhoods are young, single, and close followers (if not setters) of the latest trends and fashions. They congregate in places like Silver Lake (Los Angeles), the Pearl District (Portland, OR), and Wicker Park (Chicago). Hipsters, though, are hardly fleeing intolerance. Rather, from an urban perspective, the issue is one of gentrification; that is, the invasion of working-class neighborhoods by a more affluent group in search of affordable housing and work spaces. The subsequent rise of property values, changeover in retail activities, and transformation of the neighborhood's image often engender conflict between the newcomers and existing residents around street life, the use of public spaces, housing costs, retailing, and neighborhood politics. Intolerance arises in the conflicts that ensue as the neighborhood is transformed and existing residents feel unwelcome or, worse, are displaced.[26]

Moving away from neighborhoods and further complicating our understanding of aliens, strangers, and neighbors is that each individual and group has more than one quality that differentiates it from others. A person is not solely Jewish or solely lesbian. Identity has multiple facets. Similarly for groups. They, too, cannot be characterized in terms of a single quality such as class or age. To characterize people as Puerto Rican, evangelicals, bisexual, or liberals tells us very little about their place in the city. Moreover, the qualities that we use to distinguish people are themselves often divisible. One is Italian and, within that category, possibly Sicilian or Milanese or Neapolitan. Jews are not just Jews, but orthodox or liberal, professionals or business owners, from Canada or from Israel, assimilated or fiercely separatist. Tolerance is a matter not only of identifying one or another quality to "accept," but of accepting people for who they are as complex human beings. This requires the inclination to allow for others what one would allow for oneself—the possibility of living a full life. Tolerance rests on equanimity.[27]

In effect, the qualities on which intolerance often turns are imputed and ascribed rather than inherent to individuals and groups. At play are nationality and ethnicity, gender and sexuality, religion, race, occupation, lifestyle (for example, bohemianism), class, and even political affiliation as when a self-proclaimed, Jewish socialist displays political lawn signs in a devoutly Christian and conservative neighborhood. And while, over time, the qualities that draw attention have shifted, a number have endured as sources of difference: race, ethnicity, religion, and nationality come immediately to mind. Consider religion. Roger Williams, who founded Providence (RI), was expelled from the Massachusetts Colony in 1635 because of his criticism of the Church of England and the King. Today, Muslims who were once near-invisible (at least politically), have become an object of governmental and public anxiety after the destruction of the World Trade Center in Manhattan in 2001 by Islamic terrorists and the rise of religion-based terrorism in Europe, Asia, Africa, and the United States.

Religion is particularly important. Throughout the country's cities and over the decades, diverse religions have been more or less tolerated and the national government, despite its Christian origins, has attempted to remain independent of any one religion. The First Amendment to the US Constitution states that "Congress shall make no law respecting an establishment of religion, or prohibiting the free expression thereof." Religious tolerance seems established. In 2015, Indianapolis had nine mosques, seven synagogues, and 45 different Christian denominations including 197 Baptist churches, 44 Catholic churches, 37 Lutheran congregations, nine Episcopalian churches, 78 nondenominational bodies, and one interdenominational group.[28] All of these religions coexist more or less peacefully.

As a generalization, cities of the United States are relatively tolerant places. People of very different nationalities, religions, and sexual practices manage to live together without overt and sustained conflict. Yet, tolerance is not equally present in all cities, nor equally experienced by all groups. Some cities seem more so than others. A 2012 study created a composite measure of tolerance using the incidences of hate crimes; the diversity of the residents as indicated by the presence of same-sex couples, percent white, and percent African-American; and the state's rank on tolerance. It found that the most tolerant places were Durham (NC), Honolulu, and San Francisco. (See table 5.1.) This does not mean that prejudiced people are absent from these places, that discrimina-

Table 5.1 **Most Tolerant Cities in the United States, 2012**

Rank	City
1	Durham, NC
2	Honolulu
3	San Francisco
4	Miami
5	Baltimore
6	Seattle
7	Trenton, NJ
8	Los Angeles
9	Ann Arbor, MI
10	Austin

Source: "America's Most Tolerant Cities from San Francisco to New York," *Daily Beast*, accessed August 14, 2015, www.thedailybeast.com/galleries/2012/16/the-u-s-s-most-tolerant-cities-photo.html.

tion does not occur, or that police violence is nonexistent (as the presence of Baltimore on the list, a city with serious racial issues, attests). Within each of these cities, intolerance is certainly possible. Then, there is Trenton, a city with many poor, racial minority households. Their presence is less an indicator of the opportunities available to them there than of the lack of opportunities to advance and thus, by implication, of discrimination. That acknowledged, the study points to a background of tolerance not to the lack of intolerance.

In its ideals, the national culture is one that works against prejudices being acted on and discrimination becoming public. Support for tolerance comes from such laws as the Civil Rights Act of 1964, the Fair Housing Act of 1968, and the 1990 Americans with Disabilities Act. This legislation prohibits and punishes discrimination against African-Americans, the disabled, and women whether that discrimination occurs in the workplace or through lending practices, university admissions, or government procurement policies. The law also designates hate crimes that trigger more punitive responses than a normal assault or act of vandalism.

Tolerance, though, is experienced in many different ways and these differences are also enabled by how cities work. One of its forms is indifference; that is, an attitude that accepts the sharing of public space with others unlike oneself, but that neither validates nor rebukes their presence. This is also known as civil inattention, a "ritual regard for the other person."[29] The presence of diverse others does not engender a response one way or another. People unlike one's self are simply ignored.

Such indifference is a weak basis on which to build common bonds and alliances in the face of intolerance. It is neither one nor the other, and yet not to be wholly dismissed.

Ideally, tolerance would extend beyond indifference to provide protection and assistance to those outside one's group. People would come together to advocate for immigrant rights, as they did in many cities in 2006. Then, thousands of people rallied in San Diego, Atlanta, Dallas, Columbus (OH), and Los Angeles to protest proposed Congressional legislation that would have added to the penalties on illegal immigrants and treated undocumented immigrants and those who helped them as felons.[30] In a related vein, over 200 state and local governments have passed laws, issued executive orders, and supported nonbinding resolutions limiting the extent to which municipal government employees and agencies can assist the federal government on immigration matters. Called sanctuary laws, they have two objectives: first, to prevent immigrants from being deported to countries where they will become victims of violence and, second, to keep municipal governments separate from national immigration policies that harm city residents. There are even sanctuary cities.[31]

Making the city safer and—as a consequence—more tolerant at a much smaller scale are the security cameras and well-lit and widely visible waiting areas deployed to make bus stops and subway platforms safer for women in the evening, thereby recognizing a gender difference salient to city life.[32] As another example, citizens might pressure their elected officials to establish a commission on religious tolerance or speak out against Arab "profiling" by the police. Going even further, tolerance can be and has been extended to offering assistance to groups who might otherwise be treated intolerantly; for example, dedicating public funds to combat housing discrimination or to support gay health clinics. These actions project tolerance beyond indifference.

Tolerance is also filtered through the multitude of occupations and professions in which people are involved and the need for businesses to engage with customers and suppliers and strive to be profitable. Retailers must accept potential buyers from many different worlds if they wish to maximize sales. They will not survive if they turn away customers. Even ethnic businesses—a Salvadoran restaurant, a Russian nightclub, a college preparatory firm serving Korean high school students—that cluster in ethnic neighborhoods, are unlikely to discriminate even as they strive to occupy a market niche. Ethnic neighborhoods are never saturated with one group or immune from outsiders. Commerce seemingly encourages indifference if not a strong strain of tolerance.[33]

An interesting variation on this argument has to do with port cities. Places like San Francisco, New York City, and New Orleans are supposedly more tolerant, in part because, as major ports, they attract seafarers, businesspeople, and travelers from around the world. Trade requires interaction with strangers and thus acceptance into the city of large numbers of outsiders who are only passing through or using the city as a temporary place of business or work. San Francisco is a good illustration of this: "a natural experiment in the consequences of tolerating deviance."[34] From the late nineteenth through to the twenty-first century, it has been a port-of-call for ships from Asia, various countries in South America, and US cities extending as far as the East Coast. The Gold Rush of 1849 attracted people from around the world including sailors from Massachusetts who were abandoning the waning whaling industry. The city's residents understood that these strangers were essential to the city's economy even as they often quarantined them in certain areas and treated them shabbily. The conditions were thus set for San Francisco to attract bohemians in the 1950s and hippies in the 1970s. By the 1980s, the city had become a haven for gay men and eventually a center of political activism around gay rights. Today, hipsters are equally as prevalent.

Tolerance, though, is more than a by-product of commerce. It stems from the interdependencies that a city creates among its residents. People cannot provide for themselves and are dependent on others for medical care, food, transportation, entertainment, and information. Even if one withdraws into an ethnic enclave, buying only from co-ethnic businesses, attending the ethnic-based church or mosque, greeting only fellow ethnics on the sidewalks, speaking only the native language, and watching ethnic television, any attempt to ride public transportation or interact with a public agency (the post office) or nonprofit organization (say, for visa services or help with a difficult landlord) is going to lead into a maelstrom of differences. Deciding not to come into contact with people unlike one's self is difficult to do in a city, whether it entails purchasing a ticket from a bus driver, having dental work done, or signing a contract with a business owner to deliver cloth so that your workers can produce women's blouses. And although businesses, occupations, and professions have been known to discriminate, with cabdrivers refusing to pick up African-Americans or technology firms hiring only men in executive positions, these are exceptions. There is no escaping the city's interdependencies and thus the need for at least a modicum of tolerance toward others.

Before shifting the emphasis to intolerance, we need to acknowledge the way in which the city's spatial form contributes to tolerance. With similar people living together in the same neighborhoods, shopping in the same retail districts, and even working in the same industries, they are unlikely to come into frequent contact with those different from them. And with these groupings widely distributed across the city, people are not confronted on a daily basis with situations that might trigger intolerance. The city's residents might be densely packed together, but they are also segregated from each other in numerous ways. They are not just occupying different neighborhoods; bus and subway lines often serve distinct parts of the city and keep groups apart even as they connect them. The commuter rail line coming in from the affluent suburbs has a much different ridership than bus lines connecting low-rent, immigrant neighborhoods to the downtown core. Although people might mingle in the downtown, at professional sporting events, and at parades, they are, on the whole, more or less isolated from diverse others at work and at home where most people spend most of their time. This protects their differences but also enhances a tolerance that is less than ideal. Nevertheless, the social diversity of cities and the dependence that people have on each other in their daily lives brings people together and makes it important that differences be treated with respect.

Intolerance

Although tolerance seemingly characterizes US cities (more or less and for some groups more than others), intolerance lurks in the background, mostly unexpressed, but emerging in isolated and episodic instances. Every so often, it erupts in sustained and violent ways. Intolerance takes a number of forms ranging from deliberate indifference to marginalization and even oppression, with African-Americans, queers, Muslims, and recent immigrants disproportionately targeted. Too frequently, intolerance escalates to violence manifested in riots and hate crimes. When it becomes systemic, groups are stigmatized and segregated.[35]

One way in which intolerance is manifested is through ignoring a group that wishes to be publicly recognized; that is, by being actively indifferent as to their presence and distinctiveness. Much of daily life is a matter of disregarding the fact that you are surrounded by people who are unlike you, believe things that you consider wrong-headed, and behave in ways that you find inappropriate. These others speak a foreign language, socialize in front of their homes rather than in the backyard, have large families, are isolationist on foreign policy, or smoke cigarettes. To the extent that these differences do not affect how you live with and interact with them, they are of no concern. But this form of distancing from and avoidance of others—an attitude of "live and let live"[36]—is not itself tolerance. In its ideal form, tolerance requires that we acknowledge differences and do so non-prejudicially. The acknowledgment does not have to be public, as when the government allows a Korean neighborhood to have bilingual street signs, but it should involve the acceptance of such differences as part of what it is to be human. Differences need to be acknowledged without prejudice; that is, without morally devaluing them.

Indifference, then, has a Janus-like quality—some times acting as the basis for tolerance, other times breeding intolerance. Its dark side can be subtle as when a public school system has no course in its curriculum on the history of a group whose presence in the city is large and increasing. Or, it might manifest in a less subtle way when school administrators ignore the language problems faced by the children of immigrants and fail to hire teachers who speak their language or offer courses that might improve their English, or, when the police department rejects the request of a growing immigrant community for a liaison so that the police will have a better understanding of that community's needs and culture. An early example involves the New York City government in the late nineteenth century. It attempted to enforce standards of behavior in Central Park by prohibiting political activities along with "swimming, fishing, playing musical instruments, and posting notices or parading for civic or military purposes." Here was a clear expression of intolerance toward working-class and immigrant leisure activities and an example of middle-class indifference. More recently, public housing authorities have attempted to eliminate certain forms of lower-class behavior such as impromptu social gatherings in public spaces, thereby imposing more acceptable, middle-class norms.[37]

More blatant examples of indifference appear as acts that marginalize people, a second form of intolerance. They include denying parade permits for groups, passing laws that prohibit the homeless from occupying public spaces, and slating for development a site (for example, an African-American or native American burial ground) that a group believes is worth memorializing and protecting. In both Boston and New

York City, the organizers of the St. Patrick's Day Parade have denied permission to gay and lesbian Irish individuals to march as an identifiable group. The mayor of Birmingham in 2008 rejected a request by the Central Alabama Pride to hold a march as part of a gay and lesbian pride celebration. More recently, at least 21 cities have passed laws prohibiting the sharing of food with the homeless in public spaces and numerous cities have banned sleeping or begging there as well.[38] This marginalization might be less formalized in law when a group new to a neighborhood is discouraged by long-term residents from using a public park for activities—music and dance, picnicking, sports—they find unacceptable.

Marginalization also manifests around claims to "being an American." This mainly involves immigrants, but has also been extended to political activists. The prevailing cultural norm seems to be that newcomers should make every attempt to assimilate into American culture and that no efforts (for example, accommodating foreign language in public schools) should be mounted that hinder that assimilation. People are expected to adopt English as their primary language (particularly in public), pursue citizenship, and embrace patriotism. The goal is a nation of "Americans" different from the nations of France, Bolivia, Syria, or Thailand. These expectations are not unique to the United States. Yet, many people identify as Irish-American or Cuban-American and wish to hold onto their cultural heritage. They continue to speak the language of their birth, dress in ways that express their national origins, celebrate festivals special to them, and (for some) engage in non-Christian religious practices.

As one example, a number of cities—Hazelton (PA) and Green Bay (WI)—have passed laws declaring that English is the official language of the municipal government, thereby forbidding government employees from speaking to residents in another language (for example, when they apply for a marriage license) or from making governmental documents available in a language other than English. As one supporter of such laws noted, "Government must encourage immigrants to join mainstream society and not live in linguistic isolation." Although this might be admirable in one sense, it ignores the challenges of adapting to a new country and sends a message that immigrants are unwanted. For this reason, the mayor of Nashville in 2007 vetoed such a bill, declaring that it would make the city "less safe, less friendly, and less successful."[39]

Third, intolerance is often experienced as violence and can come from government or from individuals or groups in the form of hate crimes. Of the former, consider police violence against African-Americans. In the first seven months of 2015, 197 black people (predominately male) were killed by the police with blacks three times more likely to die in this way than whites. One-third of these killings occurred in the 100 largest cities of the country with Chicago, New York, and Los Angeles having the highest numbers of these deaths. Tulsa (OK), Hialeah (FL), and North Las Vegas are particularly dangerous cities for blacks. Racial profiling, in which police officers assume certain behaviors on the basis of racial characteristics (for example, skin color, dress, even location), is a contributing factor. Critics argue that the police have an implicit bias that views young, black males as potential criminals.[40]

All such incidents are tragic, but one that reverberated further than most occurred in Ferguson (MO) on August 9, 2014. That day, Michael Brown, a black teenager, was walking with a friend along a city street. The owner of a nearby convenience store had just called the police about a robbery and the police went in search of the suspect. Brown was killed when, as he was walking away, a police officer shot him. In response, the black community held a candle-light vigil, but not too many hours passed before outrage erupted into violent protest. Angered, in part, by what they believed to be racist treatment by the police, some of the protesters took to vandalism and looting; others engaged in peaceful demonstrations. The police responded with force including the use of tear gas and the city government had to declare a state of emergency. This went on for over a week. For the black community, it was another in too many instances where a black man had been wrongfully shot by the police.[41]

Police violence is hardly a recent occurrence. During the late 1960s in Philadelphia, the police commissioner spoke publicly of his dislike of both political demonstrators and African-American civil rights protesters. In one instance, in August 1970, and after skirmishes between the police and black youth, the police raided the offices of the radical Black Panther Party, one of a number of militant groups active in the city. The raid was intended as a preemptive tactic to cripple the organization and block its People's Revolutionary Convention scheduled for later that year. Gunfire was exchanged and, after the police used tear gas to empty the building, the Panthers were taken outside, stripped naked, and searched—a photograph of which was published in the local newspaper. Public condemnation of the police department followed.[42]

For the most part, the 1960s were a particularly violent time with the police and members of the black community clashing repeatedly and violently. The country's major cities experienced riots with almost all of them triggered by a police incident. In Detroit, the precipitating factor was a police raid of an after-hours social club where illegal drinking and gambling was underway. The police arrested 82 patrons and rumors of police violence quickly spread throughout the black community. Within hours a crowd had gathered, looting began, and the situation spun out of control. At the end, 43 people were killed, 33 of them black. In 1967, riots were particularly numerous with over 30 cities having such disturbances and approximately 90 people killed, over 2,000 injured, and another 11,000 people arrested. The great plurality of those arrested, killed, and injured were black.[43]

African-Americans are not the only ethnic minority that has faced violent acts of intolerance. A particularly notorious episode occurred in Los Angeles in June of 1943—the Zoot Suit Riots.[44] For months, the tensions between servicemen stationed in the city and young Mexican-American males had been escalating. Mexican-American young men had taken to wearing broad-shoulder jackets and balloon-leg trousers as a public statement of identity. Doing so defied a wartime ban on using wool for clothing. More to the point, their attire was a rebellion against the dominant culture and how Mexican-Americans were being treated in the United States. The soldiers and sailors viewed the Zoot-Suiters as unpatriotic. One night a group of servicemen chartered cabs and went to the Mexican-American neighborhood. There, they beat and stripped the youths and burnt the suits. Thousands of military personnel were involved with the police arresting nearly 600 Mexican-Americans, but few of the soldiers and sailors. The riots ended only with the intercession of military officials. And while a citizen's investigatory commission noted that racism was one of the causes of the riots, the then-mayor blamed the insurrection on juvenile delinquents.

The psychological damage caused by systemic discrimination and intolerance has consequences that spill out in ways that further diminish urban life. In the early 1990s, Jonathan Coleman, a journalist, went to Milwaukee (Wisconsin) to learn more about black-white relations. One of the people he interviewed was Maron Alexander. As Coleman tells it, Alexander

> was tired of hearing about the Iranians or Koreans or whoever owned the corner store. He got tired of it because some of these stores had been owned by black families but often a family's children had no interest in carrying on. So the stores would get sold, and it wouldn't be long before someone was throwing a brick through one of the windows, angry as hell that what money they had was being given to a

foreigner, to someone they believed had an easier go of things than they did. It all got jumbled up, and the anger often got misdirected, and there was no easy way to untangle it."[45]

Violence against and by minorities is not the only consequence of racial intolerance. Intolerance can also be opportunistic as when individuals or small groups set out to harm those who they find to be threatening or unacceptable. When such behavior is motivated by a bias against individuals because they belong to a specific group, it is considered a hate crime. In chapter 4, discussing neighborhood governance, I mentioned how white residents in Detroit attempted to prevent African-American families from moving into their neighborhoods by picketing the house, throwing stones at the new occupants, and blocking the moving vans. Or, to take a more contemporary example: children in public schools who are different are often singled out by their classmates for bullying.

Hate crimes include physical assault, systematic harassment, graffiti, property damage, and threatening behavior. And, although hate crimes are uncommon relative to all crimes, they nonetheless persist in their occurrence. Consider just a few, brief examples. In 2014 in Albuquerque, a man made an anti-Semitic threat against the Jewish owner of a delicatessen. Two years earlier in Atlanta, three men attacked a gay man with a tire iron as he was leaving a grocery store. That same year in Compton (CA) two men assaulted a 17-year-old African-American male with a metal pipe. Two women in Cleveland in 2009 affixed a toy camel, hung by a noose, on the door of a tenant of Arab descent in an attempt at intimidation. In Jackson (MS), "a city with a long history of racially motivated violence," seven white teens in 2011 selected a black man at random and ran over him with a truck. During the attack, which led to the man's death, one of them shouted "white power."[46]

Like most intolerance, the incidence of hate crimes varies from one city to the next. Using an index created by combining an anti-bullying score, the existence (or not) of anti-discrimination laws, and the presence (or not) of a human rights commission, 24/7 Wall Street created a list of the worst cities for lesbian/gay/bisexual/transgendered (LBGT) rights. Leading the list was Southaven (MI) followed by Irving, Lubbock, and Mesquite in Texas, and Great Falls (Montana). Alabama was well represented with three cities—Tuscaloosa, Huntsville, and Mobile—in the top ten, but it failed to dethrone Texas with four cities (add Laredo to the list above).[47] In these cities, hate crimes against queer people are more likely.

The homeless are particularly vulnerable to attack. Between 1999 and 2010, there were 1,184 documented acts of violence against homeless people by non-homeless individuals and another 353 such events in the subsequent three years. Approximately one-half of these incidences were beatings and nearly 4 out of 10 were assaults with a deadly weapon. In the earlier, eleven-year period, these acts resulted in 312 deaths. Here is a vulnerable population: stigmatized and ostracized, lacking the protection of shelter, and clustered in cities.[48]

Muslims have also been the target of hate crimes. Since the foreign terrorist attacks in the United States in September 2001, the number of such crimes has increased five-fold.[49] In 2015, an estimated 260 hate crimes were committed against them. Muslims have also been subject to public abuse and derision. One of the most infamous incidents occurred in Dearborn (MI), a city where approximately one-third of the 97,000 residents is Muslim. Muslims there have periodically faced anti-Muslim demonstrations. In 2011, several dozen people gathered to condemn the Muslim presence in America. They were met by over 600 counter-protesters. In 2014, the same group returned to demonstrate outside of a mosque. It claimed that the city was being run according to Sharia law. Anti-Muslim sentiment, generally, has been on the rise and protests against the construction of mosques have occurred across the country. In Phoenix in 2015, armed protesters gathered to denounce Islam with "the organizer of the protest [calling] it a patriotic sign of resistance against what he deemed the tyranny of Islam in America."[50]

The anti-Muslim activities of the 2000s were very much a consequence of the 2001 terrorist attacks widely viewed as a threat to national security. With the US invasions of Iraq and Afghanistan and the rise of radical Taliban and Islamic State (ISIS) groups in the Middle East and Africa, attitudes toward Muslims shifted against them. Religious intolerance was further fueled by a terrorist attack in Paris in November 2015 that killed 130 people and injured nearly 400 others, another such event in Brussels in March 2016 that left over 30 people dead and 300 injured, and a Syrian refugee crisis that seemed to portend an invasion of terrorists. (Home-bred terrorism is also part of this story.) Political rhetoric during the 2015–2016 Presidential primary campaign, particularly on the Republican side, contained much public vilification of Muslims that fueled an increase in anti-Muslim protests, acts of intimidation, and violence. One of the Republican candidates, Donald Trump, who was later elected President, called for the deportation of all Syrians from the country, the cessation of all Muslim emigration, and even the assassination of the relatives and families of terrorists. One poll indicated that 47 percent of all Americans believed that Islamic values are antithetical to the American way of life and noted deep-seated anxiety stemming from violent events as well as "heated political rhetoric."[51]

Similar concerns arose during the Second World War. Then, Americans of German descent were viewed with suspicion and those of Japanese heritage were interned in camps for the duration of the conflict. Prior to that, the Chinese Exclusion Act of 1882 (repealed in 1943) prohibited immigration from China. A fear of Communism led to the persecution of political radicals just after the first and second world wars.

One can understand the anxiety that people might have toward others different from themselves. And, while the resultant intolerance is not always triggered by living in a city, it is exacerbated there: the fear becomes more visceral through the nearness of strangers. When such groups demonstrate for their rights in the downtown where people shop or take up residence in a neighborhood or become employed in the firm where one works, the potential for intolerance rises. Proximity, and the inability to avoid it, despite segregation of homes, occupations, and industries, enables intolerance to thrive.

Intolerance, of course, does not always result in violence. It is often less physically intimidating even if harmful. No less oppressive and extending beyond the government, this fourth form of intolerance is rooted in structural injustice.[52] Consider, for example, the many ways in which groups limit access by other groups to private and public resources and opportunities, the most pervasive being discrimination in jobs and housing. Certain groups are denied opportunities so that others can live well.

People are not randomly distributed across labor markets but are rather clustered in industries and occupations with many of those industries and occupations characterized by ethnic, racial, and immigrant divisions.[53] Since the development of city police departments, many have been dominated by one or another group. In Boston in the mid-twentieth century, the Irish had a monopoly on the police force and in cities like Philadelphia and Phoenix until the 1970s police officers were almost always white. City fire departments have had a similar history, as have public school teachers. White firefighters in New York City in 2012 organized to protest a court mandate to assure that qualifying tests were designed and scored in such a way as to give minorities an equal opportunity to join the fire department. (The department was then 89 percent white when the city was approximately 44 percent white.) One firefighter said: "I feel that I'm being discriminated against because I am Caucasian."[54]

In more recent times, Mexicans have become the majority of back-of-the-house restaurant workers in Los Angeles, Hassidic Jews dominate the diamond and photographic supply industries in New York City, and Korean women predominate in nail salons. Such ethnic or racial dominance blocks other groups from taking these jobs or entering these industries. The barriers are both symbolic and self-inflicted (as when a person assumes that they will be uncomfortable or unwanted) and also formalized in tests and actively produced through outright discrimination (as when a job applicant is simply turned away because she does not fit the profile of a business's workforce or its retail clientele).

As regards occupations, groups develop ethnic niches that marginalize non-ethnic groups through a combination of intolerance toward others (for example, biased examinations or workplace harassment) and hiring practices that operate on the basis of social ties. Historically, the unionized construction trades have resisted the inclusion of African-Americans and women. In Milwaukee in 2000, 93 percent of the steel metal workers, 92 percent of electricians, and 88 percent of carpenters were Caucasian. During the two previous decades, African-Americans made hardly any impact numerically among electricians, carpenters, plumbers, or brick masons. More generally, and switching to the national level, 98 percent of all speech language pathologists and 92 percent of all dieticians and nutritionists were white in 2014. Asian presence was highest in software developers (32 percent), African-Americans in home health aides (36 percent) and barbers (36 percent), and Hispanics in drywall installers (62 percent) and agricultural graders and sorters (54 percent). Not all of this concentration is voluntary and not all is attributable to discrimination. Intolerance, though, has to be acknowledged as it works its way through education, social relations, and hiring practices.[55]

Despite a host of anti-discrimination laws, intolerance is ever-present in city housing markets. At one time in the country's history, women were not allowed to obtain a home mortgage without having a man cosign the loan. African-Americans have been blocked from buying homes by restrictive covenants placed on deeds, denial of mortgages by banks, and real estate agents who steer them away from white neighborhoods. They are also less likely to live near white households. The typical African-American household in 2010 lived in a neighborhood that was 35 percent white, whereas the typical white household lived in a neighborhood that was 75 percent white. In other words, "whites live in neighborhoods with low minority participation." Hispanics are discriminated against in homeownership and rental markets as well. Residential segregation is a "principal organizational feature" of American cities with the consequences of discrimination—involuntary segregation and isolation—visible in the urban landscape.[56]

Contributing to racial residential segregation through to the 1970s was the practice of "redlining" that began in the 1940s and still lingers. In pushing the housing industry to move to long-term, amortized mortgages so that more people could afford to become homeowners, the federal government also required banks to adhere to systematic appraisal and leading practices, especially when dealing with government-backed mortgages. This included taking into account the changing racial composition of the neighborhoods where loans were being considered. Neighborhoods where African-Americans lived and were increasing in number were considered to be of high risk of deterioration and thus to pose a high probability that home values would fall and the borrowers would default on their loans. Maps were drawn to help bankers assess these conditions with red lines used to outline the neighborhoods most at risk. Consequently, few government-backed mortgages were given to African-Americans in cities. They were forced to either take on risky financing or remain renters. In St. Louis, five times as many of these government-backed mortgages went to households in the suburbs (where whites were relocating) as went to those in the city where blacks were confined. Racial segregation deepened.[57]

Racial segregation peaked some time in the 1960s and 1970s and has declined since then, though only slightly. As measured by the percentage of the nonwhite population that would have to move for an area to be integrated (that is, to match the proportion in the larger population), African-Americans are still highly segregated from whites. In cities like Detroit and Milwaukee, nearly 8 out of 10 households would have to change neighborhoods to create a geographical balance. This is less the case for Hispanics and Asians. In Los Angeles and New York City, just over 6 out of every 10 Hispanic households would have to move and in Houston and Boston almost 5 out of every 10 Asian households would need to relocate. These cities are the most segregated as regards minorities. (See table 5.2.) Despite the improvement since the 1970s, discrimination against these groups in the housing market continues.[58]

Admittedly, other factors compel minorities to live in segregated settings. The discrimination that they suffer in public schools and labor markets produces incomes that limit their housing choices. In addi-

Table 5.2 Metropolitan Areas with Highest Levels of Racial Segregation, 2010

Black-White Segregation Index	Index	Hispanic-White Segregation Index	Index
Detroit	79.6	Los Angeles	63.4
Milwaukee	79.6	New York City	63.1
New York City	79.1	Newark, NJ	62.6
Newark, NJ	78.0	Boston	62.0
Chicago	75.9	Salinas, CA	60.9
Philadelphia	73.7	Philadelphia	58.5
Miami	73.0	Chicago	57.0
Cleveland	72.6	Oxnard, CA	54.5
St. Louis	70.6	Santa Ana, CA	54.1
Nassau-Suffolk, NY	69.2	Houston	52.5

Note: The table uses the Index of Dissimilarity. It measures the distribution of two groups across the Census tracts of a metropolitan area in comparison to the distribution of those groups in the metropolitan area as a whole.
Source: John R. Logan and Brian J. Stults, "The Persistence of Segregation in the Metropolis: New Findings from the 2010 Census," US2010 Project (New York: Russell Sage Foundation, 2011).

tion, they can face discrimination when they visit a rental office, work with a real estate agent to find a home, approach a bank to purchase a mortgage, or even, after renting an apartment or home, have difficulty convincing the landlord to provide agreed-upon services or do needed maintenance. This kind of discrimination—rooted in prejudice and intolerance not just for minorities but also for the disabled, immigrants, queer individuals, and people of different religious beliefs—is present in most city housing markets. In 2013, a real estate company in Philadelphia settled a lawsuit alleging that it had steered African-American, but not white, prospects to high crime neighborhoods. The Seattle Office of Civil Rights in 2015 filed a discrimination complaint against a rental property owner who asked African-American and Hispanic inquirers about their criminal records, while same-sex couples were given fewer applications and brochures and were showed fewer available units than heterosexual couples. And in Richmond (VA), a rental company settled a case alleging that it had discriminated against disabled tenants with wheelchairs by requiring a higher security deposit and proof of $100,000 in liability insurance. Despite an array of fair housing legislation and advocacy organizations defending open housing laws, such actions persist. They also continue to be sought out and punished.[59] Involuntary segregation denies these groups access to good homes, denies their children access to better schools, and denies the family a better location relative to work opportunities, health care, dry cleaners, and policing.[60]

The fact is that cities are characterized by the sorting of households, services, and amenities into different types of neighborhoods. This sorting is mainly driven by class or income differences, but race (African-American, Hispanic, Asian) and immigrant status are consequential as well, with African-American households particularly burdened. Housing quality and cost, the presence of good elementary schools and accessible and quality retail and personal services, the quality of local parks and playgrounds, and even the provision of public services such as street maintenance and trash collection are all more or less aligned. Food prices are often higher in poor, segregated neighborhoods and access to fresh fruits and vegetables is limited. It is not only the family in which one is born that is crucial to one's future life, but also the neighborhood in which one lives. In a country where educational credentials are frequently the pathway to a good job and subsequent advancement, the quality of the local neighborhood school can make a significant difference. And, schools are not the only issue. Access to employment and business opportunities is also related to where one lives with more affluent neighborhoods better connected than poorer ones.

For certain groups, the accessibility of resources and opportunities is also blocked in private businesses such as banks where people of color or women are denied loans for starting up businesses, public schools where expectations about performance and the lack of compensatory support result in immigrant and minority children being deprived of educational resources, and in local politics when certain groups are prevented from holding elected office due to the organization of electoral districts. African-Americans in cities have for decades complained of being snubbed by taxi drivers who fear picking them up or being denied service at a fancy restaurant or a posh shop.

Competition for resources and opportunities not only threatens the livelihood of already-disadvantaged groups, it also engenders even more insecurity about living well. The resultant tension can easily erupt in marginalization, exploitation, and even violence of one group against another. Within cities, groups cannot avoid this competition. Neighborhoods are constantly changing as residents move about in search of better (or less or more expensive) places to live. People are displaced by gentrification and reluctant (and sad) to move. Business and jobs appear and disappear, thereby disrupting employment and making people's lives more precarious. Moreover, to the extent that the capitalist economy favors flexibility over stability and prefers labor to be unorganized, peoples' lives are made uncertain.[61] Many at the low end of the educational and social spectrums are forced to work multiple, contingent jobs in order to survive. Local taxes are constantly being adjusted and proposed, public services are being reallocated across neighborhoods and groups, and electoral influence is constantly shifting. For groups to remain passive in the face of such conditions is to be diminished. If the group does not protect its gains, those gains will be eroded. The city fosters competition and that competition, in turn, is a breeding ground for intolerance.

Urbanity

One of the consequences of intolerance is its diminution of the urbanity that makes cities vibrant. Urbanity is a quality of public togetherness and its essence is that people "can act together, without the compulsion to be the same."[62] The idea of urbanity is meant to capture what it feels like to live among people different from one's self—existing within a dense environment of strangers and an entanglement of incomparable activities. Where urbanity reigns, people feel comfortable among unknown others and act without anxiety. They welcome the many differences that the city presents to them. In such an environment, tolerance is nurtured and intolerance stifled. An ideal urbanity is inclusive and free from the class-based and gender divisions that characterized it until the mid-twentieth century. Combined with democracy, it involves people in "collective action for public benefit."[63]

In recent times, elected officials, developers, architects, and planners have supported projects to create urbanity through neighborhoods of mixed retail uses, transit stops, and a vibrant street life comprising downtown plazas, waterfront parks, a mix of attractive stores, and an array of restaurants and theaters. The goal is to use this urban quality to attract tourists and lure people away from suburban living, thereby maintaining the city's prosperity. The message is that it is acceptable to "be urban." In these neighborhood and citywide public spaces, people from different backgrounds and groups mingle. They enjoy being together in public and take advantage of the opportunity to see an impromptu street performer, sip a drink at a sidewalk café, watch their children play in a fountain, or simply luxuriate in being outside with others on a beautiful day.

That said, critics claim that these spaces seem to be mainly for those with discretionary income and leisure time, suggesting that the city is being designed for a consumption-oriented middle class. Implied is that the working class are too busy just "getting by" and the lower class too geographically distant from these places to take advantage of them. Moreover, such encounters with others are ephemeral: people make contact only for a brief moment and usually only visually. Any contacts they have are centered on leisure and consumption, not politics or workplace concerns. Critics fret about turning the city, or at least parts of it, into a theme park. Intolerance and the city's general grittiness are conveniently hidden. As one urban observer has commented, "Such pseudo-city culture offers scenes of city life, not the city itself. The City Lite is safe, orderly, and simplified."[64]

Intolerance dampens and can even destroy urbanity. It discourages people from mingling in public. The result is that only people similar to each other gather there. With differences erased, one could just as well be in a small town or village where mystery and serendipity are absent. There, one does not have to reflect on the diversity of people and how they live. Certainly, socially homogeneous places (whether an ethnic enclave or gay neighborhood) have value: they nurture group solidarity and provide a respite from the city. But, they enhance urbanity only indirectly and might even stifle it. What is most bothersome is when intolerance discourages particular people from enjoying the city's urbanity. In many cities, for example, the police and business associations act to prevent the homeless and teenagers from lingering in places meant for tourists or shoppers. Where intolerance reigns, the benefits of proximity are diminished and the awareness of diverse others is stifled. Such a city becomes divided and these divisions inhibit interaction, complicate the city's interdependencies, and make it more difficult to govern.

Few US cities have become so intolerant that urbanity has disappeared. The exceptions have all been related to racial divides and were mostly true of cities in the South in the late nineteenth and early twentieth centuries, during what is known as the Jim Crow era.[65] With slavery abolished, the oppression of African-Americans took other forms: forced residential segregation, separate waiting areas at bus stations, public restrooms and drinking fountains for whites only, separate lunch counters, and distinct seating areas in movie theaters. Less formalized practices included paternalistic greetings—"boy"—of black men and the expectation that whites and blacks would not share the same sidewalk (blacks being the ones who had to step aside). These were all aspects of an intolerance that denied urbanity to all of the residents of these Southern cities. Of course, we should worry less about what effect this had on the city than about its implications for the quality of life and opportunities of those against whom intolerance was directed.

A city with a reputation for intolerance will be less desirable to those being marginalized, those who value tolerance, and businesses that depend on having an inclusive public image and serve a diverse mix of consumers and clients. Civic leaders in Atlanta, Georgia, in the 1960s launched a promotional campaign to counter that city's image as a Southern city inhospitable to African-Americans. Its slogan became "The City Too Busy to Hate," a slogan that alluded to racial divisions as well as to aspirations to be the commercial center of the New South.[66] The city's elite recognized that intolerance could repel both households and investors and that if the city were to grow, it would have to attract both. Tolerance thus became part of a publicity campaign: tolerance of racial diversity was good for Atlanta. What civic elites wanted to avoid was the reputation that had attached to Birmingham, Alabama, during the civil rights movement of the 1960s when the police used dogs and fire hoses on African-Americans to keep them from marching for racial justice. More recently, cities (Houston) and states (North Carolina) have been widely criticized for passing laws that deny to transgendered individuals the use of public bathrooms of their choice.[67] Protest has come not just from advocacy groups but also from businesses (for example, hotels) fearful of losing customers, performers who cancel concerts in protest, and businesses and professional associations unwilling to be associated with intolerance.

Similar issues arise in cities that have been the destination for large numbers of refugees and immigrants. Civic leaders in Tucson, Omaha, San Diego, and Dayton have struggled with their presence. As aliens, these new people challenge the dominant group's capacity for tolerance and pose numerous issues involving the delivery of public services (such as public education), the identity of the community, and how people behave in public spaces. Throughout the history of US cities, new groups have transformed housing, neighborhood retail areas, and playgrounds in ways that older groups did not anticipate. Houses are painted in "odd" colors and front yards are used for food gardens or to repair automobiles whereas before these yards had been merely ornamental. Parks become gathering places for extended family picnics or impromptu soccer games. Congregating on sidewalks is a way to escape crowded apartments. The city's urbanity is being redefined and, in the process, a previous group's way of life and way of using the city is threatened.

A classic example comes from the early post-World War II era when city governments were intent on demolishing slums and replacing them with new private and public housing. In Boston, Italians gathered on the stoops of their homes, their children played in the streets, and the women hung their laundry on the fire escapes to dry. Elected officials and planners viewed these behaviors as indicators of disorder.[68] Such uses of public space were associated with slums, not with urbanity. Civic leaders feared that images of these slums would come to represent the city among investors and middle-class households and discourage the former from starting businesses and the latter from taking up residence there. On the basis of expert assessments and elite preferences for middle-class living, local leaders embraced a narrow and illiberal understanding of urbanity.

In its multiple forms, then, intolerance not only is detrimental to urbanity—a defining characteristic of cities—but also has the potential to stifle commerce (thereby undermining the generation of wealth) and, by extension, to undermine the conditions for growth. As compared to such divided and conflict-ridden cities as Belfast during the three decades of conflict between loyalists and Irish nationalist that began in the late 1960s; Jerusalem, Tel Aviv, and Ramallah beset for years by violence perpetrated by the Israeli military and militants and Palestinian insurgents; or Johannesburg under the white apartheid regime that repressed blacks from 1949 to 1994, tolerance is much more common in cities of the United States than intolerance. And, yet, to acknowledge the obvious, the United States is quite violent when it comes to gun violence with 354 mass shootings in 2015 that led to the deaths of 462 people and the wounding of 1,314 others.[69] This is intolerance at its most reprehensible.

Conclusion

The city nurtures both tolerance and intolerance: either can take root and flourish—and both do. In the United States, however, tolerance seems to predominate. Still, certain cities are less accepting of strangers and aliens than others and the history of the country is filled with examples of enduring bouts of intolerance when immigrants, African-Americans, gays and lesbians, and political dissenters from the left of the ideological spectrum and the right (but more often from the left), have been marginalized and persecuted. Even today, not too many days pass before another instance of intolerance or another hate crime makes its way onto the local and national news. And, a kind of petty intolerance always seems to operate as people encounter those who are different and, in response, because of fear, ignorance, or sheer insensitivity, treat them with disrespect.

The mix of tolerance and intolerance in any one city, of course, is not solely due to the way cities work and are organized. Cities are not isolated from regional histories and national laws, political rhetoric, and various initiatives from "English only" legislation to restrictions on immigrant voting rights. An immigrant backlash in the United States—an event common throughout the country's history—is certainly behind the intolerance currently experienced by Mexicans in Dallas, Los Angeles, and Phoenix and Muslims throughout the country. And the national rise of a radical conservative movement known as the Tea Party in the early twenty-first century has contributed to the marginalization of various groups. This has included attempts to use voter identification and registration laws to deny minorities and immigrants the right to vote, harassment of women visiting health clinics for abortions, and efforts (ultimately unsuccessful) to prevent same-sex couples from being legally married. And, in 2015, Americans were still debating whether displaying the Confederate flag on the grounds of the South Carolina State House in Columbia was an affront to African-Americans whose ancestors had been held in slavery—a symbol of hate and of the racial order that once characterized the South—or a reminder of the distinctive history of this region—an object of cultural heritage.[70]

Nonetheless, a culture of tolerance persists, even if not all elected officials, sports personalities, media pundits, corporate executives, or heads of foundations speak and act within its bounds. Expounding to a wide audience, whether on television or through social media, a person runs the risk of harsh criticism from the media and others if he or she expresses views considered intolerant. Public claims that African-Americans are inferior or that those who cross the Mexican border illegally are mostly criminals are widely condemned.[71] Reinforcing these cultural tendencies, national, state, and local governments have numerous laws and bodies concerned with hate speech, tolerance for sexual preference, discrimination on the basis of age, and gender distinctions in employment decisions. In the United States, and currently, intolerance, and inequality for that matter, are generally viewed as inappropriate if not outright unacceptable.[72]

Finally, we should remember that tolerance and intolerance are not two separate things, but rather mutually entwined. The most obvious relationship is when intolerance gives rise to tolerance. When in June

2015 nine African-American worshipers were gunned down in church in Charleston, South Carolina, by a domestic, white terrorist, numerous people came together in support of and in solidarity with the victims and the group that they represented.[73] In almost every instance, those who committed such acts have been punished and the group that has been harmed has been offered support and even a public apology. To this extent, intolerance engenders and brings tolerance to the surface. It gives a meaning to tolerance which it would not otherwise have.

The more difficult relationship to grasp is that between tolerance and the appearance of intolerance. Acts of tolerance recognize the status of a group when that status is being contested. Those who believe that the lifestyles of queer individuals undermine Christian values and might even to be a threat to children are appalled when government leaders grant parade permits to them or allow them to marry. Public endorsement is an affront to them and might even exacerbate intolerance by hardening their attitudes. Doing so publicly, however, declares that such people deserve respect. In this sense, it is intolerance that is being challenged. Yet, tolerance can also engender a backlash when groups find tolerance so antithetical to their values that they react in anger. Like tolerance, intolerance draws meaning from its opposite. We know what intolerance is because of a national backdrop of tolerance. Cities provide the conditions for both. This is another one of those contradictions that gives to the city its unsettled character and amplifies the moral challenges of living in such diverse places.

SIX

Encountering Contradictions

Cities represent neither a triumphant resolution of life's needs and desires nor the erasure of the many disagreements and conflicts endemic to living together. My claim, a claim that is not widely embraced, underlies the argument of this book. The intent, though, is not to deny the city. Neither a contrarian call for a return to a rural existence nor a celebration of the benefits and blessings of suburbia lurks within these pages. Rather, the book represents my attempt to elaborate a critical and admiring perspective that stops short of idolatry. To do so, I took up the theme of contradictions.

Cities nurture the interplay of four contradictory forces: (1) the amassing and concentration of wealth and the deepening of poverty, (2) the destruction of ecosystems and the promise of sustainability, (3) the rewards of oligarchy and the seeming necessity for democratic practices and institutions, and (4) the pull toward intolerance coupled with the fostering of its opposite, tolerance. The heightened articulation of these forces is what defines cities. And although these contradictions have roots that extend to regions, nations, and the world itself, cities are the mechanism by which they are concentrated and given presence.

These contradictions are not immutable; they are both unstable and susceptible to change. Intolerance and oligarchy, for example, are always under pressure to give way to tolerance and democracy, environmental destruction to sustainability, extreme wealth to equality. Nothing is fixed. Intertwined, these contradictory impulses vary in their intensity across the country's cities.

Consider their instability. The balance between the oppositions contained within these contradictions is constantly shifting. Today, city governments are committed to the recycling of solid waste, a concern for sustainability that was essentially absent in the mid-twentieth century; despite decades of peaceful Muslim presence, intolerance against people of this faith is on the rise; and income inequality (relatively muted for the postwar decades) is once again ascendant. At the same time, we find many differences from one city to the next. With its heavy reliance on automobile usage, air conditioning, and sprawl, Phoenix is energy-hungry, while Boston makes more efficient use of its land, relies to a greater extent on mass transit, and has a more temperate climate including (due to climate change) a decreasing potential for icy-cold winters and thus lessening energy demand for winter heating. Chicago contains households of great affluence and neighborhoods of enduring poverty, while the residents of cities like Burlington (VT) live their lives more like each other, the extremes of income and wealth not so far apart. Depending on where they take root, the contradictions of urban life ebb and flow in intensity.

The four contradictions are also interconnected; they influence each other in myriad ways. With its concentration of wealthy families, corporations, and philanthropies, New York City generates public and private resources to fund educational programs for disadvantaged children, the reclamation of its coastline, support services for immigrants, and advocacy organizations that champion affordable housing. Oligarchic "growth coalitions" abate taxes on the real estate of the wealthy and direct public subsidies and expertise to the redevelopment of rail yards for corporate office buildings. A culture of tolerance attracts people of varied backgrounds and nationalities whose divergent lifestyles create enclaves of young, white professionals and immigrant neighborhoods with robust economies that offer numerous opportunities for making a living. Boston, Chicago, and Seattle could be described in similar terms. By contrast, in once-vibrant cities where heavy manufacturing had been dominant—places like Buffalo, Gary, Detroit, and Youngstown—contaminated soils, housing blight, and crumbling infrastructure discourage reinvestment, repel the middle class, and keep these cities poor. The fragility of working-class neighborhoods often engenders intolerance of racialized others. The contradictions are intricately interconnected: not four, separate dynamics—a tangled rather than a woven fabric.

Clearly, though, this is only one way to characterize cities. Alternatively, we could accept these contradictions, but interpret them as solely societal in nature and indifferent to how people are distributed across the land. This would cast doubt on the assertion that cities amplify their influence. And, it would have the effect of not just absolving the city of any complicity but also of easily leading, as it has done, to the dissolution of the notion of the city as a distinct form of human settlement. Or, we could reject altogether the approach I have taken and reduce the contradictions to the status of social problems that human ingenuity will eventually—and sooner rather than later—resolve. I am by no means declaring that my interpretation (one I share with others) constitutes the truth—that it somehow accesses and represents the reality of cities, appearances aside—and that all other perspectives are either ideological feints, myopic, or uninformed. The city is not one but many things. And, it is different things to different people. I happen to prefer this point of view: it explains much of what attracts and perplexes me about cities.

The task for the reader is to decide what to think about this argument; that is, to read critically and consider whether what I have proposed makes sense and resonates with prior experiences and current knowledge. One way to do this is to reflect on why this perspective should matter. "Of what use to me," the reader might ask, "is this way of thinking about the city? Does it matter that I think of cities as nurturing contradictions?" This is a fair question and my response constitutes almost all of this final chapter.

The premise of my argument is that only a hermit, someone living totally "off the grid" and isolated from others in a remote place, escapes the city and its influences. Everyone else is affected by the conditions emanating from these four contradictions. For city residents, and in their daily lives, the contradictions are even more palpable. The contradictions are significant also because they define the moral sphere that we all inhabit and establish the conditions under which we join the civic realm as political beings. To this extent, they deeply affect how we experience the city and live with others. I will conclude on this theme of "living with others." Before doing so, however, we need to briefly reflect on the many ways we sense and know the city around us. First up is the question posed above: how does thinking about the city as a Petri dish of contradictions matter?

Living amid Contradictions

To anyone living in the city, it is important whether wealth is distributed to a few leaving many others behind, whether the environment is being degraded or sustained, whether governance is controlled by a small group or spread across numerous publics, and whether those who are different in some meaningful way are treated with tolerance or intolerance. And because the conditions generated by these contradictions are unavoidable and consequential, where people live has a tremendous effect on how they navigate the dangers and opportunities that are posed. Consider the implications of these contradictions in practical, everyday terms.

When the wealthy dominate in housing markets, for example, the remaining households of the city are adversely affected. In the early twenty-first century, both San Francisco and New York attracted a group of very affluent people. Driven by technology firms and venture capital funds in the San Francisco Bay area and by the stability of property values and the wealth generated in the financial services sector in New York City, hyper-rich households were willing to pay whatever was required to purchase a property.[1] Many were plutocrats from Russia, Saudi Arabia, and other countries with volatile economies or unstable political regimes looking to place large sums of money where it could be safe. In Manhattan, developers responded by building high-rise, residential buildings on prime real estate with apartments selling at unprecedented and exorbitant prices ranging into the tens of millions of dollars and up to $100 million for penthouse status. These apartments were meant to be lived in for only a few weeks each year.

By bidding up the prices for housing, wealthy individuals in both cities also pulled up prices for middle-class and working-class households. With little available land for expanding the supply of middle-income and affordable housing, sale prices and rents escalated. Developers who might have once focused on middle-income housing turned their energies and capital to higher-end units that could yield much greater profits. This raised the costs of owning and renting in a major portion of the market. And while housing in out-of-the-way places and low-income communities was hardly touched by this phenomenon, these places were increasingly susceptible to gentrification. For the middle-class and in communities adjacent to prime real estate areas such as Central Park South in Manhattan and South of Market in San Francisco, anyone looking to buy or rent faced a highly competitive market and the prospects of paying more for housing than they would have had to pay in the absence of demand exerted by the hyperwealthy. Rising prices in the city's core forced many households to pay more or live in less attractive or less accessible areas of the city.

In cities where wealth is more evenly distributed, housing prices seem to be less volatile and developers less inclined to abandon the middle of the market for its top tier. Residents who do not count their wealth in the millions of dollars find homes to be more affordable. To the extent that housing is a big part of the wealth of middle-class and working-class households, these households are also able to take advantage of their investment. Additionally, and extending our gaze beyond the housing market, in cities where the gap between the wealthy and others is less, social cohesion is higher, life expectancies are longer, and people are more trusting of each other and consequently more tolerant.[2]

Similar comments can be made about the contradiction between environmental degradation and sustainability. It matters whether one lives in a "green" city like Portland, Oregon, or a "toxic" city such as Bakersfield in California.[3] Where the political commitment to sustainability has been weak, fewer restrictions have been placed on growth, the building of highways, and industry. As a result, soils have been made toxic, air polluted, noise levels left unabated, and ecosystems threatened. Residents in these cities are more likely to suffer from respiratory ailments and skin rashes. Their children's development is often stunted and the lives of all made less pleasant. For those who live next to commercial airports, industrial districts, or major highways; in neighborhoods with bus depots, trash transfer stations, waste recyclers, or low-end manufacturing firms; and in poorly heated homes where gas stoves and space heaters are used to stay warm; the health consequences are harmful and unavoidable. Less degraded, less harmful places exist, but moving to them is financially prohibitive for those struggling to live decently. And while affluent households can avoid environmental ills by locating in places less burdened by water pollution and toxic sites, they are not as able to escape the flooding of roads due to more severe rainstorms or the heat island effect exacerbated by climate change.

Most big city governments attempt to dampen if not eliminate these spatial disparities and make the whole city healthier; affluent cities such as Seattle do quite well in this regard. Environmental degradation is particularly acute in shrinking cities where government resources are severely curtailed and many residents suffer their consequences. A particularly striking example occurred in Flint, Michigan, in 2014, a city whose job and population loss and blighted housing market had given it prominent status as a city in decline. Unable to fund its water system, the city government attempted to save money by reducing the use of anti-corrosive agents. As a consequence, lead began leaching into the water from aging service lines. Moreover, the city government failed to counteract the rise of fecal coliform bacteria. The water became undrinkable: this in a city where the median household income was less than half of that of the state and where 40 percent of the residents were poor by governmental standards. Most alarming, it was months before the residents were warned, thus exposing them (and their children) to serious illnesses.[4]

The quality of environmental health in any city has a great deal to do with the capacity of government—federal, state, local—to develop and enforce regulations on industry, transportation, and, more generally, growth and to mount programs and policies that protect and enhance the environment. Where government is captured by local elites and developers and a growth coalition privileges unrestrained investment, environmental regulations are more likely to be sacrificed. Projects proposed by major businesses and investors are favored with legal exceptions, creative financing, and relaxed rules. Residents in these cities are thus triply burdened: by disparities in wealth, environmentally generated public health issues, and a government indifferent to their protestations.

How, though, is life different in a city where oligarchy reigns versus one where democratic practices are the rule rather than the exception? No city in the United States, I would suggest, is so dominated by one group such that its interests extend into the home or severely restrict daily life. We do not live in a totalitarian society. Whereas blacks in the late nineteenth and early twentieth centuries, particularly in the South, had to contend with widely held norms that denied them access to public restrooms, seats on a public bus, or voting rights, such oppressive controls are now prohibited by law. Cities, today, are less likely to be ruled by those with discriminatory intentions than by political coalitions intent on shaping the city around a vision of growth and development that favors middle- and upper-class values or pursues control over the city government as a way of expanding personal fortunes.

The impact of oligarchic governance is thus more indirect than direct. Residents of such cities find it difficult to resist (and are often only cursorily consulted) when decisions are made regarding subsidies to developers, large infrastructure projects such as light-rail lines, and shifts in tax policy that favor corporate interests. The city might look better in economic development brochures and lifestyle magazines and many residents of the city might well take advantage of the new shopping areas, waterfront parks, or transit options, but these projects are realized at a cost to democracy. Moreover, they represent a channeling of government resources—time, revenues, land—into things that benefit the affluent and the powerful rather than the disadvantaged. Schools, libraries, health clinics, and neighborhood playgrounds languish, while professional sports franchises enjoy government-subsidized arenas and developers profit from leasing government-subsidized, glass-clad office towers to corporate tenants. All cities are faced with a need to boost their tax bases, but some pursue this more aggressively than others and do so absent any significant distribution of the benefits to neighborhoods and less favored groups.

In cities where an oligarchic group is fixated on the private benefits of public largess, corruption is often one of the outcomes. This seems to be a common problem in shrinking cities—Detroit and Camden (NJ) are two examples—where the absence of growth and thus opportunities for wealth-generation lure morally challenged people into government and elected officials into illegal activities.[5] Opportunities to sell goods or provide services to the local government are directed to favored businesses or those willing to "pay to play." Elections are tightly controlled so as to deny voice to various groups and to create city councils where dissent is stifled. Street cleaning, playground maintenance, school upgrading, and trash removal occur more frequently in neighborhoods that support the regime than those that do not. For the poor, racial minorities, and immigrants, the differences between what their neighborhoods receive and the services enjoyed by more favored neighborhoods exacerbate the disparities in the quality of their daily lives.[6] That the former are being treated unequally is clear. What they can do about it is less obvious. In short, the balance of oligarchy and democracy matters.

One sees this most clearly in the case of policing and the black community. The absence of citizen police review boards and other mechanisms for democratic engagement with police departments is one factor in the antagonistic relationship between African-Americans and the police. One would expect that a government more open to democratic input and oversight would engender greater sensitivity on the part of the police and greater understanding of policing practices on the part of the community, thereby reducing the potential for violent confrontations. Not only a matter of the culture of police departments, it is also one of the democratic sensibilities of the local government and the extent to which various agencies are accountable to residents. Where accountability is lacking, disadvantaged groups suffer.[7]

Turning to the last of the contradictions, the daily balance or imbalance of tolerance and intolerance is probably the easiest to grasp—such acts are often quite public. Hate crimes against transsexuals are reported in the local newspaper and on local television news, a church congregation provides food and shelter for refugee families, mosques are vandalized, elected officials propose the deportation of undocumented workers, and white residents join their black neighbors for a nighttime vigil to protest the police harassment of black youth. A city council might vote to prohibit transgendered individuals from choosing which public bathrooms to use or subway riders might come to the defense of a Muslim being harassed. At the same time, intolerance and tolerance occur in less public ways. African-Americans, disabled individuals, and single mothers and their children face discrimination in the rental housing market, while gay couples intending to marry are denied venues or wedding services. These acts, whether seen by many or few, are real and consequential.

Tolerance seems to be more pronounced in affluent and dense cities.[8] There, people with a discretionary income and a love for the city feel comfortable exploring distant neighborhoods. Lesbian organizations are less likely to encounter resistance when attempting to open a community center or hold an event in the local park. Taxi drivers might be more likely to pick up African-Americans since their passengers are less likely to be headed to poor and dangerous areas of the city. For many groups, the difference between living in a tolerant city as opposed to one with an undercurrent of intolerance has direct, near-unavoidable, and personal effects. In many subtle and not-so-subtle ways, the contradictions nurtured by cities influence the conditions under which people carry on their daily lives as well as their long-term prospects for health, prosperity, safety, and happiness. Cities where inequalities are prominent are more than cities where some people have fewer resources and opportunities than others, poorer access to public and private services, and less of a likelihood for upward mobility. They are also cities where people are less healthy, both physically and mentally, and less satisfied with their lives. It can hardly be denied that "sources of social stress, poor social networks, low self-esteem, high rates of depression, anxiety, insecurity, the loss of a sense of control" have a fundamental impact on how city life is experienced.[9]

Negotiating the Moral Landscape

One of the most important of these consequences involves the moral choices compelled by these contradictions and thus the way in which residents occupy the city as political beings. This is the second part of my response to the question of why it matters that we think of the city in terms of contradictions.

Human fulfillment does not rest solely on the knowledge we acquire, the skills we learn, the goods and services we purchase, or the scope and depth of our intimate relationships. It also entails the joining of political communities in which people depend on each other for support.[10] Such communities can take various forms: the board of a cooperative apartment building, a safety committee at work, neighborhood associations, religious elders, a city council, or a regional planning body. They function as forums where we deliberate with others in order to make a world that sustains and nurtures those who occupy it.

Widely recognized is that, as members of a political community, people have rights and responsibilities attendant to that membership. Consider the various rights conferred on citizens of the United States: the right to a fair trial if accused of a crime, the right to choose which job to take of those that are available, the right to live in whatever neighborhood one can afford, and the right to marry a person of the same sex. These rights, and others, belong to us as citizens of the nation and contribute to our well-being in a variety of ways. Yet, I am less interested in rights than I am with responsibilities. Rights, particularly citizenship rights, imply a relationship between a government and its citizens as individuals; responsibilities, by contrast, focus attention on our relatedness to others. These responsibilities extend beyond the obligation to treat other people with civility as we go about our daily tasks. They include the obligations we have to acknowledge and act appropriately as regards the unjust, undemocratic, and indecent conditions created by the contradictions that infuse the city.[11]

When unbalanced in favor of wealth, environmental destruction, oligarchy, and intolerance, the residents and users of the city are confronted with stark choices. The first is whether to recognize and act on injustice or to withdraw from our responsibilities to others.[12] It is easy to complain about people struggling on the brink of destitution, groups using the local government to create wealth for themselves alone, and ecological settings being permanently destroyed, but then do nothing. People wishing to avoid these conditions can isolate themselves in affluent neighborhoods and sanitized downtowns. Of course, there are costs—moral costs—to indifference and withdrawal. Doing so places the individual outside the political community. She is no longer acting responsibly: in a strict sense, no longer due the support and rights that membership confers. Public cynicism, for this is what it is, weakens democratic practices and discourages joining with others to act.[13]

If a person does accept political responsibility, then they are confronted with a second set of choices. These involve answering the question of "what should be done?" In cities where no contradictions exist or where the contradictions are aligned to minimize injustice, these questions are less compelling. There, the need for many people to be responsible is not as pronounced. We have none of the former cities in the United States and too few of the latter. Rather, most urban dwellers are confronted with the consequences of these oppositions on a daily basis.

The issue at hand is structural injustice; that is, the systemic social processes that situate groups of people within conditions of dominance and deprivation, while enabling other groups to access a wide range of opportunities and resources that enhance their lives.[14] These processes are self-perpetuating and enduring and, while they change over time, they do so ever so slowly. In addition, it is difficult for people who are disadvantaged to move into advantaged groups; that is, to be upwardly mobile. Most importantly for this discussion, blame for these injustices cannot be easily attributable to one or another group. The responsibility for the processes that maintain poverty and deprive cities of much-needed investment is distributed among numerous individuals, groups, organizations, and institutions. To this extent, it is unclear whether blaming one person or a particular corporation is worth doing. What is required instead are collective efforts to resist injustice on a number of fronts simultaneously, a point to which I will return below.

Structural injustice, then, involves groups rather than single individuals and takes various forms: as exploitation, marginalization, powerlessness, cultural imperialism, and violence.[15] It is manifested in the enduring poverty of black households, the political disenfranchisement of immigrants and minorities, the steering of the poor and minorities to the worst neighborhoods, and the violence directed at gay and transgendered individuals. These activities and conditions are deeply implicated in the very nature of American society; they are not anomalies. A prime example in the United States involves the uninterrupted marginalization of Native Americans and African-Americans. These groups have been subjugated and deprived for centuries. Their members have been enslaved, lynched, and murdered and, for Native

Americans, slaughtered, forced off their lands, and confined to reservations. Today, both groups remain deprived and marginalized—though less so for African-Americans. This is a national embarrassment.

What is to be done? How is a person living in a city to act? What are the moral and political responsibilities appropriate to structural injustice? Civility should be a minimum. One should go through the day allowing others—with the exceptions of muggers, thieves, and other disruptors of the social peace—to be themselves. Even better would be sympathetic encounters that "promote sociability and cohesion by bridging and effacing differences." Such encounters would "ignite the potential for publics." But then what? "Do no harm" comes immediately to mind. Do not act in ways that perpetuate the injustices of the city's contradictions: pay your workers a living wage, treat others with respect, welcome immigrants, recycle. Respect is key; neither individuals nor institutions should humiliate people and injure their self-respect.[16]

This is still not enough. Political responsibility requires that we heed the various forms of injustice and speak out or act so as to make others aware, shame institutions to behave appropriately, and join like-minded people to form publics that support justice, democracy, and respect for others. Because the city is a collective experience, we should act with others. Our responsibilities are not unique to each of us as individuals, but rather held in common. In the face of contradictions, the rule should be to resist indifference, act decently, and share responsibility and do so, as often as possible, with as many people from as many different backgrounds as possible.

Important here is to recognize that some people and institutions bear more of the responsibility and have a greater capacity to respond to injustice than others.[17] Hedge funds, commercial banks, and the federal government with its affluent-friendly and corporation-friendly income tax policies, are more complicit in the imbalance of wealth than are newly arrived immigrants from Mexico. Similarly, the fair housing commissions of local governments have more tools to respond to discrimination in rental markets than a local church congregation. And, a large property management company can achieve greater energy savings in its apartment buildings than a few tenants acting alone. Government is particularly well-positioned to join with publics in addressing injustice; it figures centrally in the distribution of wealth and poverty, environmental sustainability, the nurturing (or not) of democracy, and constraints on intolerance.

On a household basis, the affluent have more money to devote to causes, better access to elected officials, and a greater capacity to create organizations to advocate for gun control than a working-class household with three children and both parents working full-time, and hoping, maybe this year, to replace the washer/dryer. Add to this the fact that people are entangled in multiple webs of support, obligation, and meaning. This makes it difficult to decide what to do. Although a person might favor the reduction of air pollution, he might also need to drive to work in order to stave off poverty. As another example of such fraught choices, the city council might favor transparency in its deliberations but be pressured by a major employer to provide a special tax break without announcing it to the public. At the same time, these entanglements present numerous opportunities for acting responsibly with others. The city demands that we act responsibly: it just does not tell us how to do so.

My objective is not to set out an agenda for what a moral person should do as a political actor. The values of justice, democracy, and respect are thin and provide little practical guidance. In reality, and as regards the choices that need to be made, the possibilities are endless and depend so much on the specificities of any unjust act or deplorable condition. What is important is that a person, a household, a church group, a local government, or a business association in some way resists the structural injustices emanating from the city's contradictions. Given the prevalence of injustice, doing no harm is only a beginning. After that, so much can be done: contributing to nonprofit organizations providing services to recently arrived refugees, supporting candidates for office who recognize the needs of less advantaged people, joining a protest march, showing respect for those with different religions and lifestyles, donating time at a soup kitchen, writing a blog about a misuse of eminent domain, boycotting a business that exploits undocumented immigrants, organizing the clean-up of a vacant lot.

One of the important reasons why this perspective on the city matters involves moral responsibilities. The city's contradictions must be seen through a moral lens and not viewed simply as phenomena that solely impact the functional requirements of daily life. Moreover, their effects are not temporary and cannot be waved away as anomalies. Consequently, my whole approach has been one that rejects the notion that the city is a self-healing, magnificent human accomplishment such that the problems it fosters are quickly addressed and eliminated. The "promise of progress" is not the foundation on which the argument rests. To believe in the inevitability and prevalence of progress is to lose sight of the moral challenges that confront us in the present moment. We begin to believe that the city's problems are fleeting malfunctions. They are not. They are inherent in the way that cities in multicultural, capitalist democracies function. They cannot be erased, only endlessly negotiated, thus compelling constant attention from morally responsible, politically engaged citizens.

Experiencing the City

This discussion of political and moral responsibilities points to an issue not yet considered to any great extent. It concerns how we experience and imagine the city. It should be obvious that I have presented the city as a factual reality: describing its conditions and processes, setting those conditions and processes within enduring contradictions, and treating the reader, to some extent, as a distant observer. Of course, we encounter the city in numerous other ways as well; for example, through the press of bodies on a crowded bus, the sudden jolt we receive when accidentally stepping off the sidewalk curb, and the piercing sounds of a siren as an ambulance passes on the street. The city is not simply real in a material sense and outside of us: the homes we live in, the workplaces some go to each day, the buses we ride, or the stream that runs through the neighborhood shopping district. It is additionally an emotional and sensory experience and, to this extent, has imaginative presence. It appears on television, in newspapers and movies, as part of discussions with neighbors, and in the world we constantly construct and reconstruct in our minds. Such perceptions are critical to how we act. The city is both a city of fact and a city of feeling and this has a great deal to do with what elected officials and social scientists tell us and how we respond to their declarations.[18]

Certainly the city has a physical presence. It is there in its buildings, taxicabs, and traffic signals and viscerally experienced through sight, hearing, smell, feeling, and taste. The skyline, viewed always from afar, symbolizes the contemporary city and defines its center of commercial activity. The buses that travel the streets make a distinctive sound, while the beeping of signals for pedestrian crossings, the honking of automobile horns, and the noises of bouncing basketballs and squeaking playground swings bring the city to us. We smell it: the smells of the city's trash removal vehicles, the inviting aromas emanating from the exhaust vent of the hamburger joint down the street, the odd chemical odor from a manufacturing plant, and the dampness that riders bring to a bus on a rainy day. And then there are the many ways that we feel the city: the hardness of concrete pavement, the jostling of bodies at a busy intersection, and the physical effort required to push open the revolving doors of large retail stores. Taste, too, is part of the urban experience. The hot dog purchased from a street cart and eaten in a rush while walking back to work, much like the coffee sipped earlier in the day along the same route, have a different taste than if consumed at home.

The city and the body are joined in particular ways and this has a great deal to do with what the city means to people.[19] Driving over a pothole or stepping in a puddle at a poorly drained crosswalk register the city quite physically. The body responds and our understanding is ever-so-slightly modified. Much has been written about how dissimilar people—diverse bodies—experience the city. The sense of fear that women often have when waiting at a bus shelter alone late at night, the concerns of transgendered individuals when walking in unfamiliar neighborhoods, the difficulties encountered by the physically disabled when negotiating their way at a crowded bus stop, and the problems faced by the infirm when crossing busy streets or descending the stairs to the subway all attest to the bodily experiences of the city.[20]

I do not mean to suggest that such experiences are peculiar to cities and, by implication, absent in small towns and country settings. We encounter these latter places through our senses as well, as part of our bodies. My point is rather that the city offers diverse experiences, not the waft of a light breeze and the smell of meadow flowers but the mix of truck exhaust, restaurant vents, and perspiring bodies. Neither do I wish to perpetuate the urban-rural trope that claims a clear and essential distinction, although it has been argued that the city engenders a particular psychological reaction that is calculating and defensive in response to the rigorous schedules of commuter trains and retail stores, the impersonal forces of the economy, and the incessant assault on our senses.[21]

Pertinent to my argument is the way in which the body is affected by the balance of urban contradictions. In cities where intolerance for African-Americans is pronounced, being black in the wrong place can well lead to physical harm. Where a robust democracy reigns, people are more likely to stage rallies for one or another cause, putting their bodies in public spaces and using their gestures and voices to make their concerns heard. Cities where the environment is a lesser priority often have more sites with toxic soils, more industrial pollution of both air and water, and ecosystems under greater threat. These conditions are experienced, in the worst of cases, as lead poisoning from faulty water systems, asthma from poor air quality, and mosquito-borne diseases from inadequate drainage. Cities with enduring and spatially concentrated poverty often combine environmental problems with intolerance toward the poor (and minorities), the latter confined to subpar housing, low-wage jobs, and unhealthy neighborhoods. In Detroit, Camden (NJ), and East St. Louis, where the poor are concentrated and local governments struggle to deliver quality public services, residents are often further burdened by political corruption and mismanagement and blocked from addressing the environmental conditions from which they suffer.

We also experience the city through public media, such as neighborhood and citywide newspapers, city magazines, local radio and television stations, neighborhood blogs, and social media sites.[22] These media representations merge with our experiences and opinions to create new understandings and give rise to new urban imaginaries. The city exists both out-there and through the media and what we experience in this way is just as "real" to us as when we walk to the corner store. On television, we see a family displaced by an apartment fire, corrupt politicians being escorted from the courthouse, and a traffic accident blocking access to a major highway along which commuters make their way home. Citywide newspapers report on murders, upcoming elections, professional sports teams, and the opening of a new restaurant. Neighborhood newspapers note community opposition to the taking of a community garden for a luxury apartment building and outrage at the shortened hours of a branch library. The neighborhood blogger bemoans the people who fail to pick up after their dogs, while the "oldies" radio station announces summer concerts in a nearby park. These media shape the city in our minds. As one consequence, the boundaries between reality and what is imagined become blurred.[23] What a city is and how we act within it are inseparable from the interpretations we share with others.

Social media have emerged as another technology mediating between people and the city. They play an increasing role in how people manage their lives and how city governments interact with their residents. Using a variety of web-based platforms such as Twitter, Facebook, and Foursquare, and able to easily access them with smartphones, people arrange to meet, hear about a "pop-up" event, praise a restaurant or a tourist site they just visited, check on bus service, or navigate a new neighborhood. City governments have adopted social media tools to provide residents with up-to-date information on public meetings and school closings, collect feedback on service performance, undertake surveys, and answer complaints.[24] Much like television and newspapers, these social media are another way through which people experience the city.

How we think about the city is also influenced by fictional accounts, not just the descriptive reporting of events. (Academic writings are an imperfect medium for conveying sensations and emotions.) Television sitcoms depict a life in the city for a family or for a group of friends. Movies imagine cities destroyed by one or another disaster—a devastating flood, a rampaging fire, a deadly disease, earthquake, space aliens. Always lurking nearby the city's prosperity—closer in Detroit than Seattle—are urban dystopias or, less dramatically, the drug gangs, squalid homes, and bribery as portrayed in detective stories, police procedurals, and television shows about "life on the streets."[25] For readers, the city appears in the memoirs of those who walk through its spaces, stories of loneliness and the search for love, crime novels, and the histories of immigrant families. These fictional depictions reflect the city of fact back to us in ways that resonate:

> From far away, they heard a rumbling that grew louder and louder until it reached them and the subway came thundering over their heads and screeched and slowed and came smashing to a stop. It exhaled and all the doors opened and the cold white light from inside the cars was cast down from high up above and the intercom spoke.[26]

Where the contradictions are most prominent, these stories become more vivid. Political corruption, the plight of the poor and the working class, and ethnic intolerance are common themes in this fictional world.

Our understanding of what it means to live in the city and how we should do so is filtered not just through various media but also through our interactions with family, friends, and acquaintances. We engage with others in a variety of ways and forums, sharing our experiences, and discussing what they mean and what we should do about them. In casual conversation, for example, we comment on the construction of a small apartment building in the neighborhood and the potential for gentrification, the local city councilperson's plan to upgrade the nearby park, the mayor's proposal to raise the property tax rate, and the acrid smell emanating from a local factory. Our experiences combine with what we have seen, read, or heard to produce a shared understanding. We experience the city not as individuals but collectively. The city is more than being stalled in a traffic jam or suffering through a garbage strike. It is also collectively imagined. What we think and know is always mediated by our interactions and conversations with others and by the television programs we watch, the newspapers we read, and the radio talk-shows to which we listen. The city is a collective experience, not an individual one. Whether we prefer to live in a city or not, in this neighborhood rather than that one, in a single-family house or a rental apartment are preferences we have with others because we experience and talk about the city together.

These experiences fragment and divide residents, while also establishing the basis for association and self-governance.[27] People with similar interests come together as publics and pressure elected officials to respond to their concerns. They picket neighborhood businesses that offer shoddy goods at exorbitant prices, organize a neighborhood crime watch, or convince neighbors to clean up a stream bed cluttered with old tires, bottles, and discarded shopping carts. Business leaders and elected officials also have an interest in sharing their knowledge and concerns. Corporate executives want the city council to defeat a proposed living wage law and, in order to convince council members to vote against it, they send lobbyists to city hall. Elected officials find it useful to listen to their constituents, caucus with others from their political party, and discuss plans for a new sports stadium with bankers, architects, professional sports teams, and city planners. Even in the most oligarchic of cities, the mayor and city council members have to maintain their political standing and this requires the creation of common perspectives among supporters. None of this emerges automatically from the facts of the city or occurs unmediated by the many ways we know it. The city is always imagined and what people imagine is always done with others.

At the same time, these perspectives are perpetually in flux. People and events are continually sending forth new information and modifying what people believe the city to be. The city eludes any singular depiction, while attempts to fix its meaning confront an obdurate reality. This is not to say that we have no stable understandings or that the city exists only in our minds. Rather it points to the interplay of facts on the ground (such as a broken water main) and how we give meaning to those facts as we go about our lives. If the water from the broken pipe seeps into the basements of adjacent buildings, this becomes a mutual concern of the buildings' owners. The seepage itself—the fact—is open to all kinds of interpretation involving the incompetence of the municipal water department, the age of the water mains, the importance of carrying building insurance, and who will be responsible for the damages. For most people—and not just in the instance of broken water mains—the city constantly, incessantly, casts up matters of collective responsibility and creates reasons to act collectively, whether that action takes the form of voicing complaints to the next-door neighbor or forming an organization. Our understanding of the city is imaginative, experiential, and emotional through and through. Yet, an actual city, in all of its physicality, exists independently and beyond these imaginaries. There is a real world. But because we can only know that world through our senses, through media, and through our interactions with others, it is not a world on which reasonable people all agree. As one observer has written, "all cities are palimpsests of . . . experiences and memories"—real, diverse, and open to interpretation.[28]

Living with Others

Clearly, urban dwellers do not negotiate their lives in the city alone. They do so as social and political beings entangled with other people. However, there is another way to think about the city as a collective phenomenon. Humans and their various associations—families, clans, government agencies, business firms, charitable organizations, churches, sport clubs—are neither the sole makers and shapers of it nor the sole mediators and victims of its contradictions. In fact, humans are not wholly and only responsible for these contradictions. Rather, the contradictions take form and thrive, or not, in a world that humans share with built forms (such as office towers and dams); nature's many species of birds, mammals, fish, and reptiles along with ecological systems such as wetlands; and regulatory (zoning, traffic lights), infrastructural (natural gas supply systems), and accounting (voter registration, dog licensing) technologies. The interaction among these various things gives the city its presence; this is how the city is constituted. The complex entanglements of people, buildings, technologies, and nature stabilize and destabilize the city's contradictions, enabling them to persist or not.[29]

Consider some of the ways in which nonhuman things enter into the contradictions we have been discussing. Poverty is a good example. It is not merely a matter of insufficient income. Poverty is perpetuated by substandard homes in which the poor are forced to live and the lesser quality and higher maintenance of the cars that they can afford to purchase. The working poor were particularly harmed a few years back by the home mortgage technologies that led in 2008 to an onslaught of foreclosures. Compared to more affluent households, poor households are more likely to live in flood plains and thus are more susceptible to the severe storms caused by climate change. Their appliances are less likely to be reliable, vermin more prevalent, and windows less tightly sealed. Affecting the health of the poor are the factories and warehouses, polluted streams, and highways adjacent to the neighborhoods where they live. By contrast, the rich can afford to reside far from environmental health hazards and, as a result, face fewer health-related problems and lower health care costs. The poor are also less likely to have internet access, more prone to falling into arrears on their utility payments, and more likely to be victimized by the privatization of infrastructure that leads to more restrictive and more costly services.[30] The rich enjoy technological advantages, are shielded from many environmental disasters, and are supported in their wealth by government-subsidized office towers, commuter rail lines, and redevelopment projects. The contradiction between wealth and poverty is embedded in the built forms of the landscape, public and private technologies, and variations in the vulnerability to nature.

One can find similar entanglements around the other contradictions. Environmental destructiveness is directly traceable to gas-fueled automobiles, building furnaces that burn low-grade heating oil, and the chemicals used in (and then exhausted from) dry cleaning businesses. The solution, and the path to sustainability, also relies on technologies: electric cars, recycling, solar panels, and green roofs. The environment is threatened by constructing homes and shops on previously undeveloped and soil-rich sites, thereby expanding the city's footprint. Democracy is enhanced by the availability of public spaces for demonstrations, easy-to-use voting machines, and transit systems that connect neighborhoods being overlooked with elected officials who have to be convinced to remedy the situation. Tolerance thrives when the media report on hate crimes, the police avoid the technology of racial profiling, gun laws are aggressively monitored and enforced, and school curriculum systems treat all students as equal in potential. Intolerance is nurtured by the walls that separate neighborhoods and digital social media open to those who spew hate.

These entanglements of humans, built forms, nature, and technologies are not specific to cities. They are found in less dense places as well. There, they are often less complex but can be equally helpful or destructive—think of individual septic systems on the helpful side and hydraulic fracking on the destructive side.[31] Ignoring the contribution of these entanglements to the city's contradictions runs the risk of seeing them as merely a matter of human intention and neglect. They are not and, because this is so, resisting them when they undermine justice, democracy, tolerance, and sustainability cannot focus solely on what humans do or do not do. Entanglements make it more difficult to negotiate these contradictions and are one of the main reasons that they persist. At the same time, they expand the possibilities for acting together.

Concluding Remarks

The preceding sentence returns us to the issue of "what should be done." I have said very little about this. It is not that I am pessimistic regarding the ability to manage the consequences of the city's contradictions, but rather because I recognize that an almost unlimited number and variety of actions could be undertaken to rebalance wealth and poverty, environmental destructiveness and sustainability, oligarchy and democracy, intolerance and tolerance. The possibilities are not infinite but neither are they easily enumerated. No single response can serve as a panacea. In addition, the question of who would have to act confronts us with the differing interpretations, interests, needs, and desires of the many groups who occupy and use the city. When compensatory actions are undertaken, some groups will benefit and others are likely to be ignored or harmed. These are the conditions with which we have to contend. We cannot erase the city's contradictions. In an ideal world, few people would be passive and many people would recognize their political obligations and moral duties.

The city is "a field of possibilities" and this should give us hope.[32] Although the city's contradictions might originate outside the city—the city being neither discrete nor isolated from the larger world—they are nurtured, amplified, strengthened, and resisted within it. A multitude of consequences are produced and, yet, distinct tendencies emerge. They are, though, just that—tendencies. And, they manifest differently over time and across social and geographical space. Neither a great human achievement nor a failure of civilization, the city is the product of an ever-changing negotiation that humans undertake with nature and with the technologies they have created. At its best, the city is an ambiguous achievement.

Acknowledgments

This book benefited from the critical comments of a number of smart people. Meg Holden, Nadia Mian, and Daphne Spain read all of the chapters and, on doing so, posed counter-arguments, offered overlooked examples, and generally encouraged me to make my argument clearer and stronger. Laura Wolf-Powers, Joan Fitzgerald, Siobhan Watson, and Eric Goldwyn read more selectively but no less perceptively. Two reviewers—one anonymous and one Alex Marshall—for the University of Chicago Press offered comments both global and particular, as did Timothy Mennel who, as Senior Editor, saw the potential in the book and encouraged me to make it more coherent and accessible.

The Rockefeller Foundation was critical to the book's development. It granted me a one-month residency fellowship at the Bellagio Center (Italy) in 2014. During this time, I developed the argument along with the scaffolding that weaves together the contradictions and empirical examples. An added benefit was being able to tap the collective experience and intelligence of my fellow residents.

The production and publicity staff at University of Chicago Press carried the burden of the post-writing phase and were always helpful and professional when doing so. I especially want to thank Mary Corrado for her editorial advice.

Finally, and once again, I am indebted to Debra Bilow for her support.

Notes

PREFACE

1. Respectively, the quotations are from Bruce Katz and Jennifer Bradley, *The Metropolitan Revolution* (Washington, DC: The Brookings Institution, 2013), p. 193; Edward Glaeser, *The Triumph of the City* (New York: Picador, 2011), p. 247; and Nan A. Rothschild and Diane diZerega, *The Archaeology of American Cities* (Gainesville: University Press of Florida, 2014), p. 192.
2. Henry B. F. Macfarland, "The Twentieth Century City," *American City* 5, no. 3 (1911):128–139. More generally, see Robert A. Beauregard, *Voices of Decline: The Postwar Fate of U.S. Cities* (New York: Routledge, 2003 ed.), pp. 27–44; Peter Hall, *Cities of Tomorrow* (Malden, MA: Blackwell, 2002 ed.), pp. 14–47; and Lewis Mumford, *The City in History* (New York: Harcourt, Brace & World, 1961), pp. 410–481.
3. On anti-urbanism in the United States, see Steven Conn, *Americans against the City: Anti-Urbanism in the Twentieth Century* (Oxford: Oxford University Press, 2014). On anti-urbanism in other countries, see Robert Beauregard, "Antiurbanism in the United States, England, and China," in *Fleeing the City*, ed. Michael J. Thompson (New York: Palgrave Macmillan, 2009), pp. 35–52.
4. For critiques of what has been labeled the "urban age thesis," see Neil Brenner and Christian Schmid, "The 'Urban Age' in Question," *International Journal of Urban and Regional Research* 38, no. 3 (2014):731–755 and Brendon Gleeson, "The Urban Age: Paradox and Prospect," *Urban Studies* 49, no. 5 (2012):931–943.

5. For a more sustained treatment of this perspective, see Marshall Berman, *All That Is Solid Melts into Air* (New York: Penguin, 1988). Berman writes that "to be modern is to live a life of paradox and contradiction" (p. 13).
6. Edward J. Glaeser, "Reinventing Boston: 1630–2003," *Journal of Economic Geography* 5, no. 2 (2005):119–153. The quotation is on p. 151.
7. Jane Jacobs, *The Economy of Cities* (New York: Random House, 1969), p. 86.
8. The first quoted phrase is from Leo Hollis, *Cities Are Good for You: The Genius of the Metropolis* (London: Bloomsbury, 2013), p. 6 and the second from Michael Shapiro, *Reading the Postmodern Polity* (Minneapolis: University of Minnesota Press, 1992), p. 122.
9. For a different treatment of contradictions, see Nicholas Phelps, *Sequel to Suburbia* (Cambridge, MA: MIT Press, 2015). He discusses three of them: growth vs. environmental conservation, individual accumulation vs. collective consumption, and localism vs. regional cooperation.
10. Andrea Wulf attributes the origins of this sense of interconnectedness to the famous nineteenth-century scientist Alexander von Humboldt. See her *The Invention of Nature* (London: John Murray, 2015).
11. On Marxist urban theory, see Mike Davis, *City of Quartz* (London: Verso, 1990) and Andy Merrifield, *The New Urban Question* (London: Pluto Press, 2014). As for non-Marxist scholars, see Glaeser, *Triumph of the City*; Witold Rybczynski, *Makeshift Metropolis: Ideas about Cities* (New York: Scribner, 2010); and Jon Teaford, *The Twentieth-Century American City* (Baltimore: Johns Hopkins University Press, 1986). For a conservative perspective, see Myron Magnet, *The Millennial City: A New Urban Paradigm for the 21st Century* (Chicago: Ivan R. Dee, 2000).
12. Ash Amin and Stephen Graham, "The Ordinary City," *Transactions of the Institute of British Geographers* 22, no. 4 (1997):411–429. See also Michael Storper and Allen J. Scott, "Current Debates in Urban Theory: A Critical Assessment," *Urban Studies* 53, no. 6 (2016):1114–1136.
13. This book is meant to join other books on "the city" written for a general audience. They include histories such as Lewis Mumford's *The City in History* (1961) and Jon Teaford's *The Twentieth Century American City* (1993); popular presentations such as Alan Ehrenhalt's *The Great Inversion and the Future of the American City* (2012), Edward Glaeser's *Triumph of the City* (2011), Alex Marshall's *How Cities Work* (2000), and Witold Rybczynski's *Makeshift Metropolis* (2010); and introductory texts such as Deborah Stevenson's *The City* (2013) and books on urban theory such as Phil Hubbard's *City* (2006) and Simon Parker's *Urban Theory and the Urban Experience* (2004).

CHAPTER ONE

1. Throughout this and other chapters, population data on cities have been taken from various US Bureau of the Census demographic data sets, most available online.
2. Of course, de-urbanization occurs. For a general overview as this applies to the United States and elsewhere, see Robert A. Beauregard, "Shrinking Cities," in *International Encyclopedia of Social and Behavioral Sciences*, ed. James D. Wright, 2nd ed., vol. 21 (Oxford: Elsevier, 2015), pp. 917–922.
3. Global Agenda Council on Competition, *The Competitiveness of Cities* (Geneva: World Economic Forum, 2014).
4. James Defilippis, "Why Urban Policy? On Social Justice, Urbanization, and Urban Policies," in *Urban Policy in the Time of Obama*, ed. James Defilippis (Minneapolis: University of Minnesota Press, 2016), pp. 293–301 with quotation from p. 294. See also Neil Brenner and Christian Schmid, "Towards a New Epistemology of the Urban?" *City* 19, no. 2 (2015):151–182.
5. On treating the city as having a single, defining trait, see Ash Amin and Stephen Graham, "The Ordinary City," *Transactions of the Institute of British Geographers* 22, no. 4 (1997):411–429. The quotation on the open-endedness of the urban is on page 5 in Louis Wirth, "Urbanism as a Way of Life," *American Journal of Sociology* 44, no. 1 (1938):1–24.
6. Jennifer Robinson, "Urban Geography: World Cities, or a World of Cities," *Progress in Human Geography* 29, no. 6 (2005):757–765, quotation on p. 763.
7. On agglomeration see Edward Soja, *Postmetropolis: Critical Studies of Cities and Regions* (Oxford: Blackwell, 2000), pp. 6–18 and Michael Storper and Allen J. Scott, "Current Debates in Urban Theory: A Critical Assessment," *Urban Studies* 53, no. 6 (2016):1114–1136. The Wirth article is "Urbanism as a Way of Life."
8. On imaginaries, see Andreas Huyssen, ed., *Other Cities, Other Worlds: Urban Imaginaries in a Globalizing Age* (Durham, NC: Duke University Press, 2008). The classic book on the city of the mind is Italo Calvino's novel *Invisible Cities* (New York: Harcourt, Brace, 1974). On self-awareness, see Ian Hacking, *The Social Construction of What?* (Cambridge, MA: Harvard University Press, 1999), pp. 31–32.
9. Daniel A. Bell and Avner de-Shalit, *The Spirit of Cities: Why the Identity of a City Matters in a Global Age* (Princeton, NJ: Princeton University Press, 2011).
10. Jane Jacobs, *The Death and Life of Great American Cities* (New York: Vintage, 1961), p. 16.
11. Witold Rybczynski, *Makeshift Metropolis: Ideas about Cities* (New York: Scribner, 2010), p. 57.
12. The US Bureau of the Census defines an urban area as "a densely developed territory that contains 50,000 or more people." See www.census.gov/geo/reference/gtc/gtc_urbanrural.html, accessed September 17, 2015.

13. This sensibility contributed greatly to the importance in the 1960s and 1970s of Herbert Gans's *The Urban Villagers: Group and Class in the Life of Italian-Americans* (New York: Free Press, 1962). See also Benjamin Looker, *A Nation of Neighborhoods: Imagining Cities, Communities, and Democracy in Postwar America* (Chicago: University of Chicago Press, 2015), pp. 135–164.
14. Alan Ehrenhalt, *The Great Inversion and the Future of the American City* (New York: Vintage, 2012), p. 185. On the decentered form of Los Angeles, see the chapters by Edward Soja and Allen Scott, Richard Weinstein, Christopher Jencks, and Michael Dear in Allen J. Scott and Edward Soja, eds., *The City: Los Angeles and Urban Theory at the End of the 20th Century* (Berkeley: University of California Press, 1996).
15. Brenner and Schmid, "Towards a New Epistemology of the Urban?," p. 166. See also Neil Brenner, "Introduction: Urban Theory without an Outside," in *Implosions/Explosions: Towards a Study of Planetary Urbanization*, ed. Neil Brenner (Berlin: Jovis, 2003), pp. 14–31.
16. Manuel Castells, *The Urban Question: A Marxist Approach* (Cambridge, MA: MIT Press, 1977). For a contemporary assessment, see David Harvey, *Rebel Cities: From the Right to the City to the Urban Revolution* (London: Verso, 2012).
17. See Robert A. Beauregard, *When America Became Suburban* (Minneapolis: University of Minnesota Press, 2006), especially pp. 19–39. For a historical perspective focused solely on the suburbs, see Kenneth T. Jackson's "classic" text *Crabgrass Frontier: The Suburbanization of the United States* (New York: Oxford University Press, 1985).
18. As one example, see Bruce Katz and Jennifer Bradley, *The Metropolitan Revolution: How Cities and Metropolitan Areas Are Fixing Our Broken Politics and Fragile Economy* (Washington, DC: The Brookings Institution, 2013).
19. For an historical depiction of this, see William Cronon, *Nature's Metropolis: Chicago and the Great West* (New York: W.W. Norton, 1991). In brief, cities are defined by both their concentration and extension. See Neil Brenner, "Theses on Urbanization," *Public Culture* 25, no. 3 (2013):85–114, especially pp. 102–104.
20. By privileging the city in the discussion, I am not denying its interdependence with its metropolitan region. Neither do I mean to devalue or denigrate the suburbs as does James Kunstler when he writes: "The nation's massive suburban build-out was an orgy of misspent energy and natural resources that squandered our national wealth." See his *The City in Mind* (New York: Free Press, 2001), p. xi.
21. On this point, see Gerald E. Frug and David J. Barran, *City Bound: How States Stifle Urban Innovation* (Ithaca, NY: Cornell University Press, 2008), pp. 12–52 and Paul Kantor, *The Dependent City: The Changing Political Economy of Urban America* (Glenview, IL: Scott, Foresman/Little Brown College Division, 1988), pp. 193–218.

22. Sam Bass Warner, "Slums and Skyscrapers: Urban Images, Symbols, and Ideologies," in *Cities of the Mind*, ed. Lloyd Rodwin and Robert M. Hollister (New York: Plenum Press, 1984), pp. 181–195. On the "branding" of cities, see Miriam Greenberg, *Branding New York: How a City in Crisis Was Sold to the World* (New York: Routledge, 2006).
23. Leo Hollis, *Cities Are Good for You: The Genius of the Metropolis* (London: Bloomsbury, 2013), p. 9.
24. Murray Bookchin takes the notion of contradictions much further than I do. In his *The Limits of the City* (New York: Harper Colophon Books, 1974), he writes that "since modern society is basically irrational, it should not surprise us that the city reflects and even exaggerates the social irrationalities of our time" (p. viii).
25. Manuel Castells, *The Urban Question: A Marxist Approach* (Cambridge, MA: MIT Press, 1977) and David Harvey, *The Limits of Capital* (Chicago: University of Chicago Press, 1982), pp. 373–412.
26. Andy Merrifield, *The New Urban Question* (London: Pluto Press, 2014), pp. 7–9.
27. Henri Lefebvre, *The Urban Revolution* (Minneapolis: University of Minnesota Press, 2003 [orig. 1970]), quotation on page 15. See also Neil Smith, *Uneven Development* (Oxford: Basil Blackwell, 1984).
28. Ira Katznelson, *City Trenches: Urban Politics and the Patterning of Class in the United States* (Chicago: University of Chicago Press, 1981), p. 3.
29. Of course, and as will be discussed, different groups have different experiences: African-Americans are less likely to view US society as basically tolerant and Mexicans and Muslims are rightly fearful given current anti-immigrant politics.
30. On this point, see Charles Tilly, *Durable Inequalities* (Berkeley: University of California Press, 1998) and Michael Walzer, *Spheres of Justice: A Defense of Pluralism and Equality* (New York: Basic Books, 1983).
31. M. Christine Boyer explores how city planning emerged to "diffuse the[se] contradictions of urban development." See her *Dreaming the Rational City* (Cambridge, MA: MIT Press, 1983), p. 68.
32. For one representative example, see Allen J. Scott and Michael Storper, "The Nature of Cities: The Scope and Limits of Urban Theory," *International Journal of Urban and Regional Research* 39, no. 1 (2015):1–15.

CHAPTER TWO

1. On how cities generate wealth, see Jane Jacobs, *The Economy of Cities* (New York: Random House, 1969); Jane Jacobs, *Cities and the Wealth of Nations* (New York: Vintage, 1985); Christopher Kennedy, *The Evolution of Great Cities: Urban Wealth and Economic Growth* (Toronto, CA: University of Toronto Press, 2011); and Mario Polese, *The Wealth and Poverty of Regions: Why Cities Matter* (Chicago: University of Chicago Press, 2009).

2. Of course, we need to consider the quality of these services and facilities, a point to be discussed below.
3. This claim, albeit of the nation rather than cities, can be found in Henry George's famous critique of the Gilded Age of the late nineteenth century—*Poverty and Progress* (1880). See Edward T. O'Donnell, *Henry George and the Crisis of Inequality* (New York: Columbia University Press, 2015).
4. Michael Walzer, *Spheres of Justice* (New York: Basic Books, 1983).
5. On different lifestyles, see Marc Doussard, *Degraded Work: The Struggle at the Bottom of the Labor Market* (Minneapolis: University of Minnesota Press, 2013); Emily Jane Fox, "It's Expensive Being Rich," *CNNMoney*, accessed October 29, 2015, http://money.cnn.com/2014/06/05/luxury/expensive-being-rich/; and Katie Johnston, "City's Middle-Income Base Eroding: Fewer Families Can Afford Boston, Analysis Shows," *Boston Globe*, September 7, 2015.
6. Peter Eisinger, *The Rise of the Entrepreneurial State* (Madison: University of Wisconsin Press, 1988). On the contribution of public wealth to solidarity and a sharing economy, see Duncan McLaren and Julian Agyeman, *Sharing Cities: A Case for Truly Smart and Sustainable Cities* (Cambridge, MA: MIT Press, 2015). On the importance of public goods and services to well-being, see Tony Judt, *Ill Fares the Land* (New York: Penguin Press, 2010).
7. On public Wi-Fi, see Josh Harkinson, "City Wi Fi: Fast, Cheap, and No You Can't Have It," *Mother Jones*, January 22, 2015, accessed October 29, 2015, at www.motherjones.com/print/268411; Alex Marshall, "Who Controls Fiber?" *Governing* 26, no. 7 (2013):24–25; and Olivia Quitana, "City Expands Public Wi Fi Network, Connects Community," *Daily Free Press*, accessed October 29, 2015, at www.dailyfreepress.com/2015/09/21/city-expands-public-wi-fi-network-connects-community. On the broader involvement of governments in city economies, see Richard L. Cook Benjamin, "From Waterways to Waterfronts: Public Investment for Cities, 1815–1980," in *Urban Economic Development*, ed. Richard D. Bingham and John P. Blair (Beverly Hills, CA: Sage Publications, 1984), pp. 23–45; Edward L. Glaeser, "Reinventing Boston: 1630–2003," *Journal of Economic Geography* 5, no. 2 (2005):119–153; Eric Monkkonen, *America Becomes Urban: The Development of U.S. Cities and Towns* (Berkeley: University of California Press, 1988), pp. 89–110; and Mary Lindenstein Walshak and Abraham J. Shragge, *Invention and Reinvention: The Evolution of San Diego's Innovation Economy* (Stanford, CA: Stanford Business Books, 2014).
8. See Richard Florida, *The Rise of the Creative Class* (New York: Basic Books, 2002) and Edward Glaeser, *Triumph of the City* (New York: Penguin Press, 2011), pp. 17–40.
9. Alex Marshall provides an overview of the many ways in which governments make the economy possible. See his *The Surprising Design of Market Economies* (Austin: University of Texas Press, 2014).

10. Harold Platt, *City Building in the New South: The Growth of Public Services in Houston, Texas, 1830–1910* (Philadelphia: Temple University Press, 1983).
11. On the evolution of the American economy and the role of government, see W. Elliot Brownlee, *Dynamics of Ascent: A History of the American Economy* (New York: Alfred A. Knopf, 1974) and Curtis Nettels, *The Emergence of a National Economy, 1775–1815* (New York: Holt, Rinehart and Winston, 1962). For a more contemporary perspective, see John R. Logan and Harvey Molotch, *Urban Fortunes: The Political Economy of Place* (Berkeley: University of California Press, 1987). To be avoided here is a technological (and transportation-based) determinism of urban growth; see Monkkonen, *America Becomes Urban*, pp. 162–164.
12. Paul Kantor, *The Dependent City: The Changing Political Economy of Urban America* (Glenview, IL: Scott, Foresman and Company, 1988). For a discussion of this in relation to streets, see Peter D. Norton, *Fighting Traffic: The Dawn of the Motor Age in the American City* (Cambridge, MA: MIT Press, 2008).
13. On the extraction and concentration of wealth generally, see Andrew Sayer, *Why We Can't Afford the Rich* (Bristol, UK: Polity Press, 2015). Lester Thurow argued that great wealth that makes a family one of the richest in the world often appears suddenly. See Lester Thurow, *Generating Inequality: Mechanisms of Distribution in the U.S. Economy* (New York: Basic Books, 1975). The chapter you are reading is not about the super-wealthy or those who suddenly ascend to their ranks.
14. Personal correspondence from Laura Wolf-Powers.
15. Raj Nalleri, Breda Griffith, and Shedid Yusuf, *Geography of Growth: Spatial Economies and Competitiveness* (Washington, DC: The World Bank, 2012).
16. Calculated from US Small Business Administrative data accessed April 7, 2015, at www.sba.gov/advocacy/firm-size-data.
17. Based on the listing provided at www.usbank/loations.com, accessed May 8, 2015.
18. Jacobs, *The Economy of Cities*, pp. 103–232 and Saskia Sassen, *The Global City* (Princeton, NJ: Princeton University Press, 1991). Of course, large corporations are not confined to using local capital; they have retained earnings and access to lenders and investors from across the country and the world.
19. Calculated from www.fortune.com/fortune500, accessed March 16, 2015.
20. Michael Storper, *Keys to the City: How Economics, Institutions, Social Interaction, and Politics Shape Development* (Princeton, NJ: Princeton University Press, 2013), pp. 32–51; Polese, *The Wealth and Poverty of Regions*, pp. 30–49.
21. On knowledge spillovers, see Michael Storper, *Keys to the City*, pp. 167–182; Jacobs, *The Economy of Cities*, pp. 49–84; and Edward L. Glaeser et al., "Growth in Cities," *Journal of Political Economy* 100, no. 6 (1992):1126–1152. This "centripetal" argument has to be read in relation to the

"centrifugal" forces—organizational, technological, and transportation technologies that enable goods and services to be produced beyond their markets—that also operate.

22. Glaeser, "Reinventing Boston: 1630–2003."
23. These data are from the 2013 American Community Survey accessed through American Factfinder March 18, 2015. The web address is www.factfinder.census.gov.
24. Alex Marshall, *How Cities Work: Suburbs, Sprawl, and the Roads Not Taken* (Austin: University of Texas Press, 2000), p. 210. See also pp. 133–155.
25. See, for example, Walshak and Shragge, *Invention and Reinvention.*
26. All of these forces can operate in reverse, as when cities shed people and businesses and property values decline. This is the case for shrinking cities. See Robert Beauregard, "Growth and Depopulation in the United States," in *Rebuilding America's Legacy Cities*, ed. Alan Mallach (New York: The American Assembly, 2012), pp. 1–24 and Robert Beauregard, *When America Became Suburban* (Minneapolis: University of Minnesota Press, 2006), pp. 40–69.
27. Joe R. Feagin and Robert E. Park, *The Urban Real Estate Game* (Englewood Cliffs, NJ: Prentice-Hall 1983) and Eisinger, *The Rise of the Entrepreneurial State.*
28. Despite all the benefits (such as tax abatements, loan subsidies, zoning changes, transit access) the public sector confers on private investors, the prevailing ideology nonetheless views government as an impediment to wealth creation and to the freedom of investors, property owners, and businesspeople. For an academic perspective, see Jamie Peck, *Constructions of Neoliberal Reason* (Oxford: Oxford University Press, 2010). On the small government movement, see Michael Cloud, 2016, "The Five Iron Laws of Big Government," Center for Small Government, accessed April 1, 2016, www.centerforsmallgovernment.com/fiveIronlaws.htm. On antigovernment groups see "Antigovernment Movement," Southern Poverty Law Center, accessed April 1, 2016, www.splcenter.org/fighting-hate/extremist-files/ideology/antigovernment.
29. Eugenie Birch, "Anchor Institutions in the Northeast Mega-Region," in *Revitalizing American Cities,* ed. Susan M. Wachter and K. A. Zeuli (Philadelphia: University of Pennsylvania Press, 2014), pp. 207–223.
30. Most of what has been written on this point comes from the field of historic preservation and the claim that architecture gives meaning to place and to those who experience it. See Paul Spencer Byard, *The Architecture of Additions: Design and Regulation* (New York: W.W. Norton, 1998), pp. 10–15. Catherine Ingraham blurs the distinction between private and public buildings in Catherine Ingraham, "Takings," *Future Anterior* 4, no. 2 (2010):31–39. On built form and a city's identity, see Daniel A. Bell and Avner de-Shalit, *The Spirit of Cities: Why the Identity of a City Matters in a Global Age* (Princeton, NJ: Princeton University Press, 2011).

31. From Table 1, p. 28 in Christopher E. Herbert, Daniel T. McCue, and Rocio Sanchez-Moyono, "Is Homeownership Still an Effective Means of Building Wealth for Low-Income and Minority Households? (Was It Ever?)," no. HBTL-06 (Cambridge, MA: Joint Center for Housing Studies, Harvard University, 2013).
32. US Department of Commerce, "The Geographic Concentration of High-Income Households, 2007–2011" (Washington, DC: US Bureau of the Census, 2013). The median household income at the time was $51,000.
33. This is despite the fact that Bridgeport is a relatively poor and economically struggling city.
34. Alan Berube, "All Cities Are Not Created Unequal" (Washington, DC: The Brookings Institution, 2014). See also Alan Berube and Brad McDearman, "Good Fortune, Dire Poverty, and Inequality in Baltimore: An American Story" (Washington, DC: The Brookings Institution, 2015), Figure 9.
35. Sam Roberts, "Manhattan's Income Gap Is Widest in U.S., Census Finds," *New York Times*, September 17, 2014.
36. Herbert, McCue, and Sanchez-Moyono, "Is Homeownership Still an Effective Means of Building Wealth for Low-Income and Minority Households? (Was It Ever?)," Table 4, p. 35.
37. Charles Tilly, *Durable Inequalities* (Berkeley: University of California Press, 1998).
38. Brief biographies on Gates and Bezos can be found on the Biography website, which can be accessed at www.biography.com/people/
39. Malcolm Gladwell, "Starting Over," *New Yorker* 91, no. 24 (2015):32–37. The data are on pages 34 and 35. See also Raj Chetty, Nathaniel Hendren, Patrick Kline, and Emmanuel Saez, "Where is the Land of Opportunity?," posted February 10, 2014, accessed November 3, 2015, www.wallstreetpit.com/102380-where-is-the-land-of-opportunity, and Paul Krugman, "The Death of Horatio Alger," *Nation*, January 5, 2014.
40. See David Harvey, "Class-Monopoly Rent, Finance Capital and the Urban Revolution," in *The Urbanization of Capital* (Baltimore: Johns Hopkins University Press, 1985), pp. 62–89; David Madden and Peter Marcuse, *In Defense of Housing* (London: Verso, 2016); and Peter Marcuse, "The Enclave, the Citadel, and the Ghetto," *Urban Affairs Review* 33, no. 2 (1997):228–264.
41. Brookings Institution, *State of Metropolitan America* (Washington, DC: The Brookings Institution, 2010).
42. On how legal restrictions on cities exacerbate spatial inequalities, see Gerald Frug and David J. Barron, *City Bound: How States Stifle Urban Innovation* (Ithaca, NY: Cornell University Press, 2008).
43. Robert J. Sampson, *Great American City: Chicago and the Enduring Neighborhood Effect* (Chicago: University of Chicago Press, 2012).
44. Sayer, *Why We Can't Afford the Rich*, pp. 83–96.
45. On the importance of informal institutions and networks in economic

development, see Storper, *Keys to the City*, pp. 104–138 and pp. 167–182 and Florida, *The Rise of the Creative Class*. More generally, see Mark S. Granovetter, "The Strength of Weak Ties," *American Journal of Sociology* 78, no.6 (1973):1360–1380.

46. See Peter Dreier, John Mollenkopf, and Todd Swanstrom, *Place Matters: Metropolitics for the 21st Century* (Lawrence: University of Kansas Press, 2001), pp. 56–66 (on jobs) and William Julius Wilson, *When Work Disappears* (New York: Vintage, 1997).
47. Michael Greenstone, Adam Looney, Jeremy Patashnik, and Muxin Yu, "Thirteen Economic Facts about Social Mobility and the Role of Education" (Washington, DC: The Brookings Institution, 2013), accessed May 27, 2015, www.brookings.edu/research/reports/2013/13-facts-higher-education.
48. Paul Peterson in his *City Limits* (Chicago: University of Chicago Press, 1981) famously argued that city governments should not engage in redistributive activities but rather focus their efforts on growth and development.
49. Tonya Moreno, "City Income Taxes—U.S. Cities That Levy Income Taxes," accessed May 3, 2015, www.taxes.about.com/od/statetaxes/a/City-Income-Taxes.htm.
50. Leo Hollis, *Cities Are Good for You: The Genius of the Metropolis* (London: Bloomsbury, 2013), p. 370.
51. Edward L. Glaeser, Matthew E. Kahn, and Jordan Rappaport, "Why Do the Poor Live in Cities? The Role of Public Transportation," *Journal of Urban Economics* 63 (2008):1–24, quotation on p. 7 and Elizabeth Kneebone, "The Growth and Spread of Concentrated Poverty, 2000 to 2008–2012" (Washington, DC: The Brookings Institution, 2014).
52. Brookings Institution, "State of Metropolitan America," Figure 4, p. 139.
53. Ibid.; Ryan Childs, "These Are the Poorest Cities in America," *Time*, November 14, 2014, and Ryan Childs, "These Are the Wealthiest Cities in America," *Time*, October 30, 2014.
54. Matthew Desmond, *Evicted: Poverty and Profit in the American City* (New York: Crown Publishers, 2016).
55. Brookings Institution, "State of Metropolitan America," Figure 5, p. 141.
56. Bianca Carragan, 2014, "Being Poor in Beverly Hills, the Most Unequal City in California," posted March 6, 2014, accessed October 29, 2015, http://la.curbed.com/2014/3/6/10135416/being-poor-in-beverly-hills-the-most-unequal-city-in-california; Alexandra Cawthorne, "Elderly Poverty: The Challenge before Us," posted July 30, 2008, accessed October 29, 2015, https://www.americanprogress.org/issues/poverty/reports/2008/07/30/4690/elderly-poverty-the-challenge-before-us/; and Suzanne Travers, 2015, "Aging in New York: City Wrestles with Poverty among Seniors," *City Limits*, posted June 25, 2015, accessed October 29, 2015, www.citylimits.org/2015/06/25/nyc-wrestles-with-poverty-among-seniors/.

57. John Joel Roberts, 2014, "Where Is the Homeless Capital of America?" accessed October 29, 2015, www.huffingtonpost.com/joel-john-roberts/who-is-the-homeless-capit b 4886379.html, and US Department of Housing and Urban Development, "The 2014 Annual Homeless Assessment Report to Congress" (Washington, DC: Department of Housing and Urban Development, 2014).
58. See the literature on welfare states for discussions of how national governments influence the balance of wealth and poverty. A good beginning is Gøsta Esping-Anderson, ed., *Welfare States in Transition: National Adaptations to Global Economics* (London: Sage, 1996). On cities, see Dreier, Mollenkopf, and Swanstrom, *Place Matters*, pp. 133–172.
59. Quentin Fottrell, "The Best U.S. Cities to Grow Old In—Even If You're Poor," *MarketWatch*, posted April 12, 2016, accessed October 31, 2016, at www.marketwatch.com/story/low-income-americans-live-longer-in-these-cities-2016-04-12.
60. Nelson D. Schwartz, "Poorest Areas Have Missed Out on Boons of Recovery, Study Finds," *New York Times*, February 25, 2016.
61. David Leonhardt, "Middle-Class Blacks in Poor Neighborhoods," *New York Times*, June 25, 2015.
62. Substandard schools, a paucity of civic and religious organizations, and residential segregation are all associated with a lack of upward mobility. See David Leonhardt, Amanda Cox, and Claire Cain Miller, "Change of Address Offers a Pathway out of Poverty," *New York Times*, May 4, 2015.
63. For an early but still useful reflection on income inequality and place, see Wilbur R. Thompson, *A Preface to Urban Economics* (Baltimore: Johns Hopkins University Press, 1968), pp. 105–132. For a more recent perspective, see William W. Goldsmith and Edward J. Blakeley, *Separate Societies: Poverty and Inequality in U.S. Cities* (Philadelphia: Temple University Press, 1992).
64. Neil Smith, "Toward a Theory of Gentrification: A Back to the City Movement by Capital," *Journal of the American Planning Association* 45, no. 4 (1979):538–548.
65. Adam Gopnik, "Naked Cities: The Death and Life of Urban America," *New Yorker* 91, no. 30 (2015):80–85, quotation on page 80.

CHAPTER THREE

1. Joan Fitzgerald, *Emerald Cities: Urban Sustainability and Economic Development* (New York: Oxford University Press, 2010), p. 11.
2. See Vishaan Chakrabarti, *A Country of Cities: A Manifesto for Urban America* (New York: Metropolis Books, 2013), p. 75. In their *Sustainable Urban Metabolism* (Cambridge, MA: MIT Press, 2013), Paulo Ferrao and John E. Fernandez label cities "the cornerstone of sustainability" (p. 101). Leo Hollis writes as regards climate change that "cities might be the only solution we

have." See his *Cities Are Good for You: The Genius of the Metropolis* (London: Bloomsbury, 2013), p. 355. Lastly, see David Owen, "Greenest Places in the U.S.? It's Not Where You Think," posted October 26, 2009, accessed January, 29, 2016, http://e360.yale.edu/features/greenest_places_in_US_its_not_where_you_think/2203.

3. Ernest J. Yanarella and Richard S. Levine, *The City as a Fulcrum of Global Sustainability* (London: Anthem Press, 2011).
4. World Bank, Poverty Overview, accessed January 2, 2016, www.worldbank.org/en/topic/poverty/overview.
5. On the need to erase the divide between culture and nature, see Bruno Latour, *We Have Never Been Modern* (Cambridge, MA: Harvard University Press, 1993), pp. 97–109. On cities as socio-ecological systems, see Matthew Gandy, "Cyborg Urbanization: Complexity and Monstrosity in the Contemporary City," *International Journal of Urban and Regional Research* 29, no. 1 (2005):26–49 and Eric Swyngedouw, "Metabolic Urbanization: The Making of Cyborg Cities," in *The Nature of Cities*, ed. Nik Heyman et al. (Abdington, UK: Routledge, 2006), pp. 21–40. This argument is compatible with many of those that have emerged from the environmental movement. See Mark Roseland, "Dimensions of the Eco-City," *Cities* 14, no. 1 (1997):197–202.
6. On the presence of birds, plants, and animals and the importance of weather and geological conditions for a city, see Jennifer Wolch, Stephanie Pincetl, and Laura Pulido, "Urban Nature and the Nature of Urbanism," *From Chicago to L.A.: Making Sense of Urban Theory*, ed. Michael Dear (Thousand Oak, CA: Sage Publications, 2002), pp. 369–402 and Dawn Day Biehler, *Pests in the City: Flies, Bedbugs, Cockroaches, and Rats* (Seattle: University of Washington Press, 2013).
7. Matthew Gandy, *Concrete and Clay: Reworking Nature in New York City* (Cambridge, MA: MIT Press, 2002).
8. Richard A. Walker, *The Country in the City: The Greening of the San Francisco Bay Area* (Seattle: University of Washington Press, 2007). See also William Cronon, *Nature's Metropolis: Chicago and the Great West* (New York: W.W. Norton, 1991).
9. Walker, *The Country in the City*, p. 254.
10. Urban political ecology provides a powerful interpretive perspective with which to explore the relationship among humans, nature, and technologies and to grasp that relationship's inherently political nature. See Roger Keil, "Urban Political Ecology," *Urban Geography* 24, no. 8 (2003):723–738 and Anna Zimmer, "Urban Political Ecology: Theoretical Concepts, Challenges, and Suggested Future Directions," *Erdkunde* 64, no. 4 (2010):343–354.
11. Jane Jacobs, *The Economy of Cities* (New York: Random House, 1969), p. 3. It is the city that develops the countryside, not the countryside (by pro-

ducing a surplus) that gives rise to the city. Walker, in *The Country in the City*, remarks that "city and countryside develop in tandem" (p. 35).

12. J. R. McNeill, *Something New under the Sun: An Environmental History of the Twentieth Century* (New York: W. W. Norton, 2000).
13. Ferrao and Fernandez, *Sustainable Urban Metabolism*, p. 116.
14. My description of early Jamestown is mainly drawn from Carl Bridenbaugh, *Jamestown 1544–1699* (New York: Oxford University Press, 1980). Also see Ronald L. Heinemann, John G. Kolp, Anthony S. Parent Jr., and William G. Shade, *Old Dominion, New Commonwealth: A History of Virginia 1607–2007* (Charlottesville: University of Virginia Press, 2007) and James R. Perry, *The Formation of a Society on Virginia's Eastern Shore, 1615–1655* (Chapel Hill: University of North Carolina Press, 1990).
15. The first quotation is from Bridenbaugh, *Jamestown 1544–1699*, p. 45 and the second quotation from Jill Lepore, *The Story of America* (Princeton, NJ: Princeton University Press, 2012). The relevant chapter in Lepore is "Here He Lyes" (pp. 17–30) with the quotation on p. 21.
16. Quoted in Bridenbaugh, *Jamestown 1544–1699*, p. 50.
17. Ibid., p. 135.
18. The data are from "2000 Census: U.S. Cities over 50,000: Ranked by 2000 Density," accessed June 23, 2015, www.demographia.com/db-2000city50kdens.html. The densest metropolitan area in 2010 was Los Angeles at 2,702 people/square mile. See Wendell Cox, "New Urban Area Data Released," posted March 26, 2012, accessed June 23, 2015, http://www.newgeography.com/content/002747-new-us-urban-area-data-released.
19. McNeill, *Something New under the Sun*, pp. 269–295.
20. Quoted in Joel A. Tarr, ed., *Devastation and Renewal: An Environmental History of Pittsburgh and Its Region* (Pittsburgh: University of Pittsburgh Press, 2003), p. 3. Slag is a waste product from mining.
21. My description of Pittsburgh's environment during this era is drawn mainly from Edward K. Muller and Joel A. Tarr, "The Interaction of Natural and Built Environments in the Pittsburgh Landscape," in *Devastation and Renewal*, ed. Tarr, pp. 11–40 and Joel A. Tarr, "The Metabolism of the Industrial City," *Journal of Urban History* 28, no. 5 (2002):511–545. See also McNeill, *Something New under the Sun*, pp. 69–70.
22. Tarr, "The Metabolism of the Industrial City," p. 524.
23. William W. Buzbee, *Fighting Westway: Environmental Law, Citizen Activism, and the Regulatory War That Transformed New York City* (Ithaca, NY: Cornell University Press, 2014).
24. Robert D. McFadden, "Wall Collapses onto a Busy Manhattan Highway," *New York Times*, May 13, 2005.
25. Semiconductor production makes extensive use of electricity and water and generates hazardous gases. The water and gases then have to be

treated and disposed. See Christopher Ketty and Jason Holden, "The Environmental Impact of the Manufacturing of Semiconductors," accessed June 23, 2015, www.cnx.org/contents/ef6dfcl6–351e-42c6–9950-efc3k54cb79f@3The-Environmental-Impact-of-the-Manufacturing–of-Semiconductors.

26. Andrew Ross, *Bird on Fire: Lessons from the World's Least Sustainable City* (New York: Oxford University Press, 2011). See also Mike Davis, "Las Vegas versus Nature," *Dead Cities* (New York: New Press, 2002), pp. 85–105.
27. Ross, *Bird on Fire*, p. 244. Of course, whereas Phoenix uses a great deal of energy in the summer, it uses much less in the winter compared to cities in the colder parts of the country.
28. Ross, *Bird on Fire*, p. 246.
29. Mike Davis, *Ecology of Fear: L.A. and the Imagination of Disaster* (New York: Metropolitan Books, 1999). On the 1994 earthquake, see pp. 30–32, 37. The quotation is on p. 9.
30. Abel Wolman, "The Metabolism of Cities," *Scientific American* 213, no. 2 (1965):178–188, 190.
31. Christopher Kennedy et al., "Energy and Material Flows of Megacities," *Proceedings of the National Academy of Sciences* 112, no. 19 (2015):5985–5990.
32. Christopher Kennedy et al., "The Changing Metabolism of Cities," *Journal of Industrial Ecology* 11, no. 2 (2007):43–59.
33. The points about buildings and transportation and increasing metabolism levels come from Peter Baccini, "A City's Metabolism: Towards the Sustainable Development of Urban Systems," *Journal of Urban Technology* 4, no. 2 (1997):27–39. The excessive energy use of US cities is often connected to the availability of "cheap" fuels, especially when compared to countries in Europe.
34. David A. Theobold, "Landscape Patterns of Exurban Growth in the USA from 1980 to 2020," *Ecology and Society* 10, no. 1 (2005):32–66.
35. The vehicle ownership rate per 1,000 people was 0.11 in 1900, 222.8 in 1950, and 828.0 in 2009. See Stacey C. Davis et al., *Transportation Energy Data Book* (Washington, DC: US Department of Energy, 2014), pp. 3-5 and 3-9, and Tables 3.3 and 3.5.
36. The population data are from the US Bureau of the Census and the urban land use data from Robert A. Beauregard, *When America Became Suburban* (Minneapolis: University of Minnesota Press, 2006), p. 93, Figure 9.
37. "Cities Built on Fertile Lands Affect Climate Change," accessed April 7, 2016, www.nasa.gov/home/hqnews/2004/feb/HQ_04059_fertile_lands.html.
38. The household data are from www.census.gov/population/socdemo/hhfam/tabHH-6.pdf, accessed July 8, 2015 and the home size data are from www.money.cnn.com/2014/06/04/real estate/american home size/, accessed July 8, 2015.

39. The first quote is from Morten Daugaard, "Sprawl," in *Encyclopedia of Urban Studies*, ed. Ray Hutchinson (Los Angeles, CA: Sage Publications, 2010), vol. 2, p. 766. On the factors leading to sprawl, see Beauregard, *When America Became Suburban*, pp. 47–48, 94 and Anthony Downs, *New Visions for Metropolitan America* (Washington, DC: The Brookings Institution, 1994), pp. 6–7. The final quotation is from Daniel Lazare, *America's Undeclared War: What's Killing Our Cities and How We Can Stop It* (New York: Harcourt, 2001), p. 276.
40. Reid Ewing and Shima Hamidid, *Measuring Sprawl 2014* (Washington, DC: Smart Growth America, 2014).
41. See Alex Krieger, "The Costs—and Benefits?—of Sprawl," *Harvard Design Magazine*, Fall 2003/Winter 2004, pp. 50–55; Downs, *New Visions for Metropolitan America*, pp. 7–15; and Peter Dreier, John Mollenkopf, and Todd Swanstrom, *Place Matters: Metropolitics for the Twenty-first Century* (Lawrence: University Press of Kansas, 2001), pp. 56–91.
42. Krieger, "The Costs—and Benefits?—of Sprawl," p. 52. See also Thad Williamson, David Imbroscio, and Gar Alperovitz, *Making a Place for Community: Local Democracy in a Global Age* (New York: Routledge, 2002), pp. 71–99.
43. Robert Bruegmann, *Sprawl: A Compact History* (Chicago: University of Chicago Press, 2006), p. 17.
44. Adam Rome, *The Bulldozer in the Countryside: Suburban Sprawl and the Rise of American Environmentalism* (Cambridge: Cambridge University Press, 2001).
45. See Ferrao and Fernandez, *Sustainable Urban Metabolism*, pp. 68, 81–82 and William Rees and Mathis Wackernagel, "Urban Ecological Footprints: Why Cities Cannot Be Sustainable—and Why They Are a Key to Sustainability," *Environmental Impact Assessment Review* 16 (1996):223–248. The ecological footprint is the spatial version of urban metabolism.
46. The prior quotation is from David Moore, *Ecological Footprint Analysis: San Francisco-Oakland-Fremont, CA* (Oakland, CA: Global Footprint Network, 2011), p. 3 and the quotation in this sentence from McNeill, *Something New under the Sun*, p. 289.
47. The ranking of countries is from Mathis Wackernagel et al., "The Ecological Footprint of Cities and Regions: Comparing Resource Availability with Resource Demand," *Environment & Urbanization* 18, no. 1 (2006):103–112, Table 1, p. 108. The data on megacities is from Christopher A. Kennedy et al., "Energy and Material Flows of Megacities," *Proceedings of the National Academy of Sciences* 112, no. 19 (2015):5985–5990.
48. Such studies are quite sensitive to the quality of the available data, the assumptions on which the analysis is predicated, and the chosen methodology. Consequently, numerous inconsistencies exist from one study to the next. See Moore, *Ecological Footprint Analysis*.
49. "Energy Use in Cities," accessed November 8, 2016, www.theworldisurban.com/2011/03/energy-use-in-cities.

50. Alex Marshall, *How Cities Work: Suburbs, Sprawl, and the Roads Not Taken* (Austin: University of Texas Press, 2000), p. 175. See also Edward Glaeser, *Triumph of the City* (New York: Penguin Press, 2011), pp. 199–222. For a more critical perspective, see Michael Neuman, "The Compact City Fallacy," *Journal of Planning Education and Research* 25, no. 1 (2005):11–26.
51. Chakrabarti, *A Country of Cities*, p. 81. It could also be argued that the tall buildings that provide a dominant image for the city embody more energy in their steel and concrete and glass than buildings in the suburbs where wood-framed homes are more prevalent.
52. On "green" transportation, see Susan M. Opp and Jeffrey L. Osgood Jr., *Local Economic Development and the Environment* (Boca Raton, FL: CRC Press, 2013), pp. 67–92.
53. Sadhu Aufochs et al., *The Guide to Greening Cities* (Washington, DC: Island Press, 2013), pp. 39–48. The quotation is from Fitzgerald, *Emerald Cities*, p. 78.
54. Fitzgerald, *Emerald Cities*, pp. 40–42.
55. Sarah Goodyear, "The City of Philadelphia Wants You to Stop Ignoring Icky Overflowing Sewers," *Next City*, posted October 17, 2014, accessed April 7, 2016, https://nextcity.org/daily/entry/are-cities-ready-to-embrace-the-culture-of-green-infrastructure.
56. Chakrabarti, *A Country of Cities*, pp. 98–99. Neuman, "The Compact City Fallacy," argues for more emphasis on process than urban form and notes that the size and scale of a city might be more important than density and its morphology.
57. See Tom Daniels, "Smart Growth: A New American Approach to Regional Planning," *Planning Practice and Research* 16, no. 3–4 (2001):271–279 and Anthony Downs, "Smart Growth: Why We Discuss It More than We Do It," *Journal of the American Planning Association* 71, no. 4 (2005):367–378.
58. On the importance of density to the viability of downtowns, see Marshall, *How Cities Work*, pp. 6–17.
59. Peter Calthorpe, *The Next American Metropolis: Community and the American Dream* (New York: Princeton Architectural Press, 1993).
60. Minimizing building energy use is particularly important. See "2016 Energy Star: Top Cities," accessed April 7, 2016, www.energystar.gov/buildings/topcities.
61. Aufochs, *The Guide to Greening Cities*; Cynthia Rosenzweig et al., "Cities Lead the Way in Climate Change Action," *Nature* 467 (2010):909–911; and Stephen M. Wheeler, "State and Municipal Climate Change Plans," *Journal of the American Planning Association* 74, no. 4 (2008):481–496.
62. Rebecca Leonard, "Green Infrastructure Grows Up," *Planning* 81, no. 6 (2015):16–21.
63. Rebecca Tuhus-Dubrow, "L.A. Existential," *Slate*, posted April 19, 2015, accessed July 1, 2015, http://www.slate.com/articles/news_and_politics/

metropolis/2015/04/the_third_los_angeles_can_it_truly_become_a_green_sustainable_city.html. These changes, of course, are supported by state and federal policies. On the Los Angeles River, see Matthew Gandy, *The Fabric of Space: Water, Modernity, and the Urban Imagination* (Cambridge, MA: MIT Press, 2014), pp. 145–183.

64. See Witold Rybczynski, *Makeshift Metropolis: Ideas about Cities* (New York: Scribner, 2010), pp. 181–188. As a contrary position, consider the claim that "the only modern communities that are remotely sustainable at present are some aboriginal communities that have existed for centuries." See Roseland, "Dimensions of the Eco-City," p. 201.
65. Ebenezer Howard, *Garden Cities of Tomorrow* (Cambridge, MA: MIT Press, 1965 [orig. 1898]).
66. For an introduction to and critique of such efforts, see Mike Hodson and Simon Marvin, "Urbanism in the Anthropocene: Ecological Urbanism or Premium Ecological Enclaves?" *City* 14, no. 3 (2010):299–313.
67. Duncan McLaren and Julian Agyeman, *Sharing Cities: A Case for Truly Smart and Sustainable Cities* (Cambridge, MA: MIT Press, 2015), p. 88.
68. For food, and as regards energy consumption, what matters is its production, not its distribution.
69. John Light, "Cities Leading the Way in Sustainability," posted January 4, 2013, accessed July 1, 2015, http://billmoyers.com/content/12-cities-leading-the-way-in-sustainability/.
70. Wheeler, "State and Municipal Climate Change Plans."
71. Glaeser, *Triumph of the City*, p. 222.
72. Eric Sharp, "City Lots Become Wildlife Habitats," *Detroit Free Press*, October 16, 2008.
73. Terry Schwarz, "The Cleveland Land Lab: Experiments for a City in Transition," in *Cities Growing Smaller*, ed. Steve Rugare and Terry Schwarz (Cleveland, OH: Cleveland Urban Design Collaborative, 2008), pp. 71–83 and Joseph Schilling and Raksha Vasuderan, "The Promise of Sustainability Planning for Regenerating Older Cities," in *The City after Abandonment*, ed. Margaret Dewar and June Manning Thomas (Philadelphia: University of Pennsylvania Press, 2013), pp. 244–267.
74. Simin Davoudi, "Resilience: A Bridging Concept or a Dead End," *Planning Theory & Practice* 13, no. 2 (2012):299–307; Susan Fainstein, "Resilience and Justice," *International Journal of Urban and Regional Research* 39, no. 1 (2015):157–167; Danny MacKinnon and Kate Driscoll Derickson, "From Resilience to Resourcefulness: A Critique of Resilience Policy and Activism," *Progress in Human Geography* 37, no. 2 (2012):253–270, and Lawrence J. Vale and Thomas J. Campanella, eds., *The Resilient City: How Modern Cities Recover from Disaster* (Oxford: Oxford University Press, 2005).
75. See *Detroit Future City: 2012 Detroit Strategic Framework Plan*, accessed July 8, 2015, www.detroitfuturecity.com.
76. On the need for a sustainable city to be a democratic city, see Robert A.

Beauregard, "Democracy, Storytelling, and the Sustainable City," in *Story and Sustainability: Planning, Practice, and Possibility for American Cities*, ed. Barbara Eckstein and James Throgmorton (Cambridge, MA: MIT Press, 2003), pp. 65–77.

CHAPTER FOUR

1. See Bernard Crick, *Democracy: A Very Short Introduction* (Oxford: Oxford University Press, 2002), p. 13 and Sara M. Evans and Harry C. Boyte, *Free Spaces: The Sources of Democratic Change in America* (New York: Harper and Row, 1986), pp. 3–5.
2. Thanks to Meg Holden for this point. See Richard Sennett, *The Uses of Disorde*r (New York: W.W. Norton, 1970).
3. Benjamin R. Barber, *If Mayors Ruled the World: Dysfunctional Nations, Rising Cities* (New Haven, CT: Yale University Press, 2013), p. 53.
4. The defining text of the period, and the one most highly associated with the call for democracy, was Lincoln Steffens' *The Shame of the Cities*, published in 1904.
5. During those years, the District of Columbia was effectively governed by a Congressional committee and, unlike the states, did not have elected representatives at the federal level. As for its growth, like many older cities in the United States, Washington's population dipped significantly after 1950 but began to expand again in the 2000s. It peaked at 802,000 residents in 1950, fell to 572,000 in 2000, and increased thereafter to an estimated 658,000 in 2014.
6. See Dennis R. Judd and Todd Swanstrom, *City Politics: Private Power & Public Policy* (New York: Harper Collins, 1994), pp. 53–75. These are sometimes called growth regimes and are different from caretaker regimes committed to small government and anti-growth regimes having an environmental and/or anti-poverty agenda.
7. See Craig Calhoun, "Civil Society and the Public Sphere," *Public Culture* 5, no. 2 (1993):267–280; Jeffrey C. Goldfarb, *Civility and Subversion: The Intellectual in Democratic Society* (Cambridge: Cambridge University Press, 1998), pp. 78–102; Iris Marion Young, "Civil Society and Social Change," *Theoria* no. 83/84 (1994):73–94; and Iris Marion Young, *Inclusion and Democracy* (Oxford: Oxford University Press, 2000).
8. See Robert N. Bellah et al., *Habits of the Heart: Individualism and Commitment in American Life* (New York: Harper & Row, 1986), pp. 167–249 and Robert D. Putnam, *Bowling Alone: The Collapse and Revival of American Community* (New York: Simon & Schuster, 2000).
9. Russell Shorto, *The Island at the Center of the World* (New York: Vintage, 2005), p. 3.
10. Eric H. Monkkonen, *America Becomes Urban: The Development of U.S. Cities*

and Towns 1780–1980 (Berkeley: University of California Press, 1988), pp. 89–110; Judd and Swanstrom, *City Politics*, pp. 42–47.

11. The most famous commentator on associations in the United States is Alexis de Tocqueville. In his *Democracy in America* he writes that "the citizen of the United States is taught from infancy to rely on his own exertions in order to resist the evils and the difficulties of life." Alexis de Tocqueville, *Democracy in America*, vol. 1 (New York: Vintage Books, 1990 [orig. 1835]), p. 191.
12. Eric Fure-Slocum, *Contesting the Postwar City: Working-Class and Growth Politics in 1940s Milwaukee* (Cambridge: Cambridge University Press, 2013).
13. Chris Tilly, "Next Steps for the Living Wage Movement," *Dollars & Sense*, posted September 1, 2001, accessed July 28, 2015, http://dollarsandsense.org/archives/2001/0901tily.html.
14. See Laura Wolf-Powers, "Community Benefits Agreements and Local Government: A Review of Recent Evidence," *Journal of the American Planning Association* 76, no. 2 (2010):141–159 and Laura Wolf-Powers, "Community Benefits Agreements in a Value Capture Context," posted 2012, accessed November 17, 2016, www.works.bepress.com/laura_wolf_powers.
15. Greater Miami Chamber of Commerce website, accessed July 29, 2015, www.miamichamber.com. Better Business Bureaus are another form of business association. They are mainly concerned with regulating businesses by mediating between consumer complaints and individual businesses.
16. Alexandria Chamber of Commerce website, accessed July 29, 2015, www.alexchamber.com.
17. Sharon Zukin, *The Culture of Cities* (Cambridge, MA: Blackwell Publishers, 1995), pp. 24–38 and Lawrence O. Houston Jr., ed., *Business Improvement Districts* (Washington, DC: Urban Land Institute, 2003).
18. Evan McKenzie, *Privatopia: Homeowner Associations and the Rise of Residential Private Governments* (New Haven, CT: Yale University Press, 1994).
19. Thomas W. Sanchez, Robert E. Lang, and Dawn M. Shavale, "Security vs. Status? A First Look at the Census's Gated Community Poll," *Journal of Planning Education and Research* 24, no. 3 (2015):281–291.
20. "Trayvon Martin Shooting Fast Facts," CNN Library, accessed April 13, 2016 www.cnn.com/2013/06/05/us/trayvon-martin-shooting-fast-facts and Black Lives Matter website, accessed March 24, 2016, www.blacklivesmatter.com/about.
21. Samuel Zipp, *Manhattan Projects: The Rise and Fall of Urban Renewal in New York* (New York: Oxford University Press, 2010), pp. 197–249 and Christopher Klemek, *The Transatlantic Collapse of Urban Renewal: Postwar Urbanism from New York to Berlin* (Chicago: University of Chicago Press, 2011), pp. 129–201.
22. Jane Jacobs, *The Death and Life of Great American Cities* (New York: Vintage,

1962), pp. 112–140 and Klemek, *The Transatlantic Collapse of Urban Renewal*, pp. 187–191.

23. Mandi Isaacs Jackson, *Model City Blues: Urban Space and Organized Resistance in New Haven* (Philadelphia: Temple University Press, 2008).
24. Nicholas Lemann, *The Promised Land: The Black Migration and How It Changed America* (New York: Alfred A. Knopf, 1991), pp. 111–221.
25. H. Briavel Holcomb and Robert A. Beauregard, *Revitalizing Cities* (Washington, DC: Association of American Geographers, 1981), pp. 46–50; James DeFilippis, *Unmaking Goliath: Community Control in the Face of Global Capitalism* (New York: Routledge, 2004); Neil R. Pierce and Carol F. Steinbach, *Corrective Capitalism: The Rise of America's Community Development Corporations* (New York: The Ford Foundation, 1987); Thad Williamson, David Imbroscio, and Gar Alperovitz, *Making a Place for Community: Local Democracy in a Global Era* (New York: Routledge, 2002), pp. 213–235.
26. Dudley Street Neighborhood Initiative website, accessed July 30, 2015, www.dsni.org/dsni-historic-timeline.
27. Beryl Satter, *Family Properties: How the Struggle over Race and Real Estate Transformed Chicago and Urban America* (New York: Metropolitan Books, 2009).
28. Thomas J. Sugrue, *The Origins of the Urban Crisis: Race and Inequality in Postwar Detroit* (Princeton, NJ: Princeton University Press, 1996). The quotation is on p. 255 and the phrase "defensive localism" is on p. 210.
29. Kimberley Kinder, *DIY Detroit: Making Do in a City without Services* (Minneapolis: University of Minnesota Press, 2016).
30. Great Nonprofits website, accessed July 29, 2015, www.greatnonprofits.org/city/Denver/CO.
31. Evans and Boyte, *Free Spaces*, pp. 85–101 on the Temperance Movement and pp. 55–61 on Rosa Parks.
32. On women's organizations, see Daphne Spain, *Constructive Feminism: Women's Spaces and Women's Rights in the American City* (Ithaca, NY: Cornell University Press, 2016) and Daphne Spain, *How Women Saved the City* (Minneapolis: University of Minnesota Press, 2001).
33. Relevant here are small and declining voter participation rates, and not just in cities. See Anna Orso, "Philly's voter turnout low, but not the worst," posted November 3, 2015, accessed April 11, 2016, www.billypenn.com/2015/11/03/how-low-can-turnout-go-how-philly-ranks-with-other-big-cities-in-mayoral-elections/.
34. In Philadelphia, in the early to mid-twentieth century, the district tax assessor who determined the value of a homeowner's or businessperson's property and thus the tax bill was also the precinct captain for the political party in power. In the late 1940s, however, reformers displaced the political machine that supported such irregularities. See Joseph S. Clark and Dennis J. Clark, "Rally and Relapse: 1946–1968," in *Philadelphia: A*

300-Year History, ed. Russell F. Weigley (New York: W. W. Norton, 1982), pp. 650–657.

35. For a brief description of the Tweed Ring in New York City and its resistance to progressive politics, see Edward T. O'Donnell, *Henry George and the Crisis of Inequality* (New York: Columbia University Press, 2015).
36. Barbara Ferman, *Challenge to Growth Machines: Neighborhood Politics in Chicago and Pittsburgh* (Lawrence: University Press of Kansas, 1996), pp. 54–63; Larry Bennett, "The Mayor among His Peers: Interpreting Richard M. Daley," in *The City Revisited: Urban Theory for Chicago, Los Angeles, New York*, ed. Dennis R. Judd and Dick Simpson (Minneapolis: University of Minnesota Press, 2011), pp. 242–272.
37. Erastus Corning II of Albany (NY) might hold the record: 1942–1983.
38. On the mayor of Harrisburg, see Marc Levy and Mark Scolfero, "Ex-Mayor of Pennsylvania Capital Faces Corruption Charges Related to Failed Wild West Museum," *U.S. News and World Report*, July 14, 2015. On urban corruption, see Jim Dwyer, "Corruption in New York: An Unscrupulous History," *New York Times*, April 20, 2016. Foreshadowing chapter 5 on tolerance and intolerance, "one measure of civility is the degree to which force and fraud are kept at bay." See Jim Sleeper, "Boodling, Bigotry, and Cosmopolitanism," *Dissent* 34 (1987):413–419. The quotation is on p. 417.
39. Peter Marcuse, "New York City's Community Boards: Neighborhood Policy and Its Results," in *Neighbourhood Policies and Programmes*, ed. N. Carmen (London: Macmillan, 1990), pp. 145–163 and Tom Angotti, *New York for Sale: Community Planning Confronts Global Real Estate* (Cambridge, MA: MIT Press, 2008).
40. Austin website, accessed August 1, 2015, www.austintexas.gov. On political participation generally, see Putnam, *Bowling Alone*, pp. 31–47.
41. David Harvey, "From Managerialism to Entrepreneurialism: The Transformation in Urban Governance in Late Capitalism," *Geografiska Annaler* 71, no. 1 (1989):3–17.
42. Clarence Stone, *Regime Politics: Governing Atlanta, 1946–1988* (Lawrence: University Press of Kansas, 1989).
43. Robert A. Beauregard, "Resident Hiring Preference Ordinances: A Comparative Analysis," *Economic Development Quarterly* 1, no. 2 (1987):124–135.
44. Alan Altshuler and David Luboff, *Mega-Projects: The Changing Politics of Urban Public Investment* (Washington, DC: The Brookings Institution, 2003) and Judd and Swanstrom, *City Politics*, pp. 335–366. On specific cities, see Gregory T. Crowley, *The Politics of Place: Contentious Urban Redevelopment in Pittsburgh* (Pittsburgh: University of Pittsburgh Press, 2005) and K. Sabeel Rahman, 2016, "The Key to Making Economic Development More Equitable Is Making It More Democratic," *Nation*, accessed November 17, 2016, www.thenation.com/article/the-key-to-making-economic-development-more-equitable-is-making-it-more-democratic.
45. Heywood T. Sanders, *Convention Center Follies: Politics, Power, and Public*

Investment in American Cities (Philadelphia: University of Pennsylvania Press, 2014), pp. 341–429.

46. Mark Gottdeiner, Claudia C. Collins, and David R. Pickens, *Las Vegas: The Social Production of an All-American City* (Malden, MA: Blackwell Publishers, 1999).
47. Mary Lindenstein Walshak and Abraham J. Shragge, *Invention and Reinvention: The Evolution of San Diego's Innovation Economy* (Stanford, CA: Stanford Business Books, 2014).
48. Mark S. Rosentraub, *Major League Winners: Using Sports and Cultural Centers as Tools for Economic Development* (Boca Raton, FL: CRC Press, 2010).
49. Katherine Q. Seelye, "Boston's Bid for Summer Olympics Terminated," *New York Times*, July 27, 2015.
50. David Osbourne, "Reinventing Government," *Public Productivity and Management Review* 16, no. 4 (1993):349–356 and Robert A. Beauregard, "Private-Public Partnerships as Historical Chameleons: The Case of the United States," in *Partnerships in Urban Governance*, ed. Jon Pierre (London: Macmillan, 1998), pp. 52–70.
51. See Williamson et al., *Making a Place for Community*, pp. 146–164.
52. Mary Hammon, "Data-Driven: Leveraging the Potential of Big Data for Planning," *Planning* 81, no. 4 (2015):23–29 and Alex Marshall, "Big Data," *Metropolis* 33, no. 7 (2014):76–80, 86–87, 91.
53. Recognizing the multiplicity of political authorities involved in urban governance is what Warren Magnusson labels "seeing like a city." See his *Politics of Urbanism: Seeing Like a City* (London: Routledge, 2011).
54. Judd and Swanstrom, *City Politics*, pp. 107–126.
55. The federal government mainly relies on individual income and corporate taxes while state governments mainly rely on individual and corporate income taxes, sales taxes, and, much less so, property taxes. See Justin M. Ross, "A Primer on State and Local Tax Policy," Mercatus Center, George Mason University, Arlington (VA), accessed August 3, 2015, at http://mercatus.org/states/default/files/Ross_PrimerTaxPolicy_v2.pdf.
56. David E. Wildasin, "Intergovernmental Transfers to Local Governments," Figure 1, posted 2009, accessed August 2, 2015, www.davidwildasin.us/wp/wildasin.intergovernmentaltransfers.pdf. On unfunded mandates, see Gerald E. Frug and David J. Barron, *City Bound: How States Stifle Urban Innovation* (Ithaca, NY: Cornell University Press, 2008), pp. 92–98.
57. Gerald E. Frug, *City Making: Building Communities without Building Walls* (Princeton, NJ: Princeton University Press, 1999), pp. 45–53.
58. These state and federal dependencies should not obscure the fact that city governments are often sources of innovation, for example, around minimum wage laws and tolerance for lesbian, gay, and transgendered people. See Claire Cain Miller, "Liberals Turn to Cities to Pass Laws Others Won't," *New York Times*, January 26, 2016.

59. Joseph F. C. DiMento and Cliff Ellis, *Changing Lanes: Visions and Histories of Urban Freeways* (Cambridge, MA: MIT Press, 2013), pp. 220–230.
60. Roger Biles, *The Fate of Cities: Urban America and the Federal Government, 1945–2000* (Lawrence: University Press of Kansas, 2011); Mark Gelfand, *A Nation of Cities: The Federal Government and Urban America 1933–1965* (New York: Oxford University Press, 1975); Robert A. Beauregard, "National [Urban] Policy," in *The Oxford Encyclopedia of American Political and Legal History*, ed. Donald T. Critchlow and Philip R. VanderMeer (Oxford: Oxford University Press, 2012), pp. 344–348.
61. Alex Marshall, *How Cities Work: Suburbs, Sprawl, and the Roads Not Taken* (Austin: University of Texas Press, 2000), pp. 157–185 and Scott Bollens, "State Growth Management," *Journal of the American Planning Association* 58, no.4 (1992):454–466.
62. Gail Radford, *The Rise of the Public Authority: Statebuilding and Economic Development in Twentieth-Century America* (Chicago: University of Chicago Press, 2013).
63. C. Ross Stephens and Nelson Wikstrom, "Trends in Special Districts," *State and Local Government Review* 30, no. 2 (1998):129–138.
64. Saskia Sassen, *Cities in a World Economy* (Thousand Oaks, CA: Pine Forge Press, 1994) and Marcus Doel and Phil Hubbard, "Taking World Cities Literally: Marketing the City in a Global Space of Flows," *City* 6, no. 3 (2002):351–368.
65. Pierre Clavel, *Activists in City Hall: The Progressive Response to the Reagan Era in Boston and Chicago* (Ithaca, NY: Cornell University Press, 2010) and Richard Edward DeLeon, *Left Coast City: Progressive Politics in San Francisco, 1975–1991* (Lawrence: University Press of Kansas, 1992).
66. Max Weber, "The Nature of the City," in *Classic Essays on the Culture of Cities*, ed. Richard Sennett (New York: Appleton-Century-Crofts, 1969 [orig. 1905]), pp. 23–46.

CHAPTER FIVE

1. Michael Walzer, *On Toleration* (New Haven, CT: Yale University Press, 1997), p. xii. Of course, people might also leave the city to exchange the isolation there for "community" in a small town.
2. Daniel A. Bell and Avner de-Shalit, 2011, *The Spirit of Cities: Why the Identity of a City Matters in a Global Age* (Princeton, NJ: Princeton University Press, 2011), p. 196.
3. I will use "queer" to refer to those who are lesbian, gay, bisexual, transgendered, and questioning (LGBTQ).
4. Morgan Lee and Jeremy Weber, "Here's What Supreme Court Says about Same-Sex Marriage and Religious Freedom," *Christianity Today*, posted June 26, 2015, accessed August 12, 2015, www.christianitytoday/

gleanings/2015/june/supreme-court-states-cant-ban-same-sex-mariage .html, and Adam Liptak, "Supreme Court Ruling Makes Same-Sex Marriage a Right Nationwide," *New York Times*, June 26, 2015.

5. Robert A. Beauregard and Anna Bounds, "Urban Citizenship," in *Democracy, Citizenship and the Global City*, ed. Engin F. Isin (London: Routledge, 2000), pp. 243–256 and Don Mitchell, *The Right to the City: Social Justice and the Fight for Public Space* (New York: Guilford Press, 2003).
6. On the necessity of publics, see John Dewey, *The Public and Its Discontents* (Athens, OH: Swallow Press, 1954 [orig. 1927]).
7. Lyn H. Lofland, *The Public Realm: Exploring the City's Quintessential Social Territory* (New York: Aldine de Gruyter, 1998), p. 237. On the variation of tolerance across cities, see Bell and de-Shalit, *The Spirit of Cities*.
8. David Kirby, "Holding Hands without Making Waves," *New York Times*, January 25, 1998.
9. See, for example, Louis Wirth, "Urbanism as a Way of Life," *American Journal of Sociology* 44, no. 1 (1938):1–24. On the differences across which tolerance is negotiated, see Walzer, *On Toleration*, pp. 52–82.
10. Catherine Fennell, *Last Project Standing: Civics and Sympathy in Post-Welfare Chicago* (Minneapolis: University of Minnesota Press, 2015), p. 7 and p. 142 respectively. On the weakness of sympathy in engaging assimilated otherness, see Gary Bridge, *Reason in the City of Difference: Pragmatism, Community Action and Contemporary Urbanism* (New York: Routledge, 2005), p. 91.
11. David A. Karp, Gregory P. Stone, and William C. Yoels, *Being Urban: A Sociology of City Life* (New York: Praeger, 1991) and Fran Tonkiss, *Space, the City, and Social Theory* (Cambridge: Polity Press, 2005), pp. 8–29.
12. For an insightful investigation of such conditions in the housing market of Chicago, see Fennell, *Last Project Standing*.
13. See, respectively, Building One Community website, accessed August 12, 2015, at www.neighborslinkstamford.org; International Neighbors website, accessed August 12, 2015, at www.international-neighbors.org; and Immigrant Pathways Colorado website, accessed August 18, 2015, at www .connectingimmigrants.org. On Pittsburgh, see Luke Nozicka, "Welcoming New Neighbors: Mayor Peduto Releases Plan to Diversify Pittsburgh," *Pittsburgh Post-Gazette*, June 23, 2015; Dylan Scott, "Immigrant-Friendly Cities Want What Arizona Doesn't," *Governing* 25, no. 12 (2012):44–50; and Robert D. King, "Why American Cities Are Fighting to Attract Immigrants," *Atlantic Monthly* 279, no. 4 (2015):55–64.
14. Lofland, *The Public Realm*, pp. 167–168.
15. Margaret Kohn, "Public Space in the Progressive Era," in *Justice and the American Metropolis*, ed. Clarissa Rile Hayward and Todd Swanstrom (Minneapolis: University of Minnesota Press, 2011), pp. 81–101 and Duncan McLaren and Julian Agyeman, *Sharing Cities: A Case for Truly Smart and Sustainable Cities* (Cambridge, MA: MIT Press, 2015). More generally on public space and tolerance, see Richard Sennett, *The Fall of Public Man*

(New York: W. W. Norton, 1992) and Richard Sennett, *The Uses of Disorder* (New York: Alfred A. Knopf, 1970).

16. Sharon LaFraniere, Sarah Cohen, and Richard A. Oppel, "How Often Do Mass Shootings Occur? On Average, Every Day, Records Show," *New York Times*, December 2, 2015.
17. Lofland, *The Public Realm*, pp. 10–11 and Ash Amin, "Animated Spaces," *Public Culture* 27, no. 2 (2015):239–257.
18. Regarding how neighbors treat each other, see Nancy L. Rosenblum's insightful *Good Neighbors: The Democracy of Everyday Life in America* (Princeton, NJ: Princeton University Press, 2016).On the elusive concept of community, see Thomas Bender, *Community and Social Change in America* (Baltimore: Johns Hopkins University Press, 1978) and Tonkiss, *Space, the City and Social Theory*, pp. 8–29.
19. A good example of this can be found in J. Anthony Lukas, *Common Ground: A Turbulent Decade in the Lives of Three American Families* (New York: Vintage, 1986).
20. Isabel Wilkerson, *The Warmth of Other Suns: The Epic Story of America's Great Migration* (New York: Vintage, 2010), p. 486.
21. "Why Do US Police Keep Killing Black Men?" BBC News, May 26, 2015, accessed August 19, 2015, www.bbc.com/news/world-us-canada-32740523.
22. See Frogtown Neighborhood Association website, accessed August 21, 2015, at www.frogtownmn.org/history and Andersonville Chamber of Commerce website, accessed August 21, 2015, at www.andersonville.org/the-neighborhood/history.
23. Christine Stansell, *American Moderns: Bohemian New York and the Creation of a New Century* (New York: Metropolitan Books, 2000), p. 68. See also Herbert Gold, *Bohemia: Digging the Roots of the Cool* (New York: Simon and Schuster, 1993) and Richard Lloyd, "Bohemian," in *Encyclopedia of Urban Studies*, ed. Ray Hutchinson, vol. 1 (Los Angeles: Sage Publications, 2010), pp. 79–81.
24. Edmund Gaither and Arnold Rampersad, "Harlem Renaissance," in *The Encyclopedia of New York City*, ed. Kenneth T. Jackson (New Haven, CT: Yale University Press, 1995), pp. 526–527; David Levering Lewis, *When Harlem Was in Vogue* (New York: Alfred A. Knopf, 1981) and Bonnie Menes Kahn, *Cosmopolitan Culture: The Gilt-Edge Dream of a Tolerant City* (New York: Atheneum, 1987), pp. 247–270 with the quotation in the previous sentence on p. 527.
25. For a general discussion of these issues, see John D'Emilio, *Sexual Politics, Sexual Communities: The Making of a Homosexual Minority in the United States, 1940–1970* (Chicago: University of Chicago Press, 1983) and Didier Eribon, *Insult and the Making of the Gay Self* (Durham, NC: Duke University Press, 2004), pp. 18–23, quote on p. 21. On the Supreme Court decision, see Robert Barnes, "Supreme Court Rules Gay Couples Nationwide Have a Right to Marry," *Washington Post*, June 26, 2015.

26. On gentrification, see Loretta Lees, Tom Slater, and Elvin Wyly, *Gentrification* (London: Routledge, 2008). For a list of hipster neighborhoods, see "America's Best Hipster Neighborhoods," accessed April 22, 2016, https://www.forbes.com/sites/morganbrennan/2012/09/20/americas-hippest-hipster-neighborhoods/#71c48438cb38.
27. On living together in difference, see Iris Marion Young, *Justice and the Politics of Difference* (Princeton, NJ: Princeton University Press, 1990), pp. 226–256. As applied to specific cities, see Kahn, *Cosmopolitan Culture*.
28. See http://www.indystar.com/story/news/crime/2016/02/29/islamic-society-draws-support-other-faith-groups-after-vandalism-mosque/81111318/, www.maven.co.il/synagogues/C3311Y3420RX, and Church Angel website, accessed August 19, 2015, at www.churchangel.com/WEBIN/indy.htm.
29. Karp, Stone, and Yoels, *Being Urban*, pp. 92–94. The quotation is in Bridge, *Reason in the City of Difference*, p. 91.
30. "United States Immigration Reform Protests," accessed August 19, 2015, www.en.wikipedia.org/wiki/2006.
31. Michael Pearson, "What's a 'Sanctuary City' and Why Should We Care?" CNN, posted July 6, 205, accessed August 17, 2015, www.cnn.com/2015/07/06/us/san-francisco-killing-sanctuary-cities.
32. Tonkiss, *Space, the City and Social Theory*, pp. 94–112 and Gerda R. Wekerle, "Women's Rights to the City: Gendered Spaces of a Pluralistic Citizenship," in *Democracy, Citizenship and the Global City*, ed. Engin F. Isin, pp. 203–217. Surveillance cameras can also be used to invade people's privacy and harass those deemed undesirable.
33. This argument for tolerance appears in claims to the unprejudiced nature of free market economies. It goes as follows: in a competitive market in which buyers and sellers are numerous and free from restrictions, business owners intent on maximizing sales and profits and under pressure from other businesses which might lure their customers away, will serve all who can pay the price for the commodity regardless of race, gender, nationality, or sexual orientation. Clearly, this does not always happen.
34. Howard S. Becker and Irving Louis Horowitz, "The Culture of Civility," *Culture and Civility in San Francisco*, ed. Howard S. Becker (New Brunswick, NJ: Transaction Books, 1971), pp. 4–19.
35. See Young, *Justice and the Politics of Difference*, pp. 39–66 on the five faces of oppression: exploitation, marginalization, powerlessness, cultural imperialism, and violence. On respect, see Avishai Margalit, *The Decent Society* (Cambridge, MA: Harvard University Press, 1996).
36. Rosenblum, *Good Neighbors*, pp. 131–149.
37. Kohn, "Public Space in the Progressive Era," p. 83. On public housing, see Lawrence J. Vale, *Purging the Poorest: Public Housing and the Design Politics of Twice-Cleared Communities* (Chicago: University of Chicago Press, 2013), pp. 280–282.

38. New Civil Rights website, accessed August 19, 2015, www.thenewcivilrightsmovement.com. On the homeless and public space, see Mitchell, *The Right to the City*, pp. 161–226 and on the ban against feeding the homeless in public see Robbie Couch, "21 U.S. Cities Outlawed Feeding the Hungry due to 'Myths' about Homelessness: Report," accessed August 19, 2015, www.huffingtonpost.com/2014/10/21/american-cities-outlawing-food-sharing_n_6021796.html.
39. Robert D. King, "Should English Be the Law?" *Atlantic Monthly* 279, no. 4 (1997):55–64. Such legislation is popular at the state level with over 30 states having passed Official English laws. See Tony Fauro, "American Cities Debate English-Only Legislation," posted 2009, accessed August 21, 2015, www.citymayors.com/society/us-english-only.html for the quotation.
40. See Mapping Police Violence website, accessed August 21, 2015, www.mappingpoliceviolence.org. And, "Why Do Police Keep Killing Black Men?" BBC News.
41. Emily Brown, "Timeline: Brown Shooting in Ferguson, Missouri," *USA Today*, accessed August 21, 2015, www.usatoday.com/story/news/nation/2014/08/14/michael-brown-ferguson-missouri-timeline/14051827/.
42. "Headquarter Raids," *Philadelphia Tribune*, September 1, 1970, accessed January 20, 2016, http://wordpress.com/spotlight/headquarters-raids/.
43. See Robert A. Beauregard, *Voices of Decline: The Postwar Fate of U.S. Cities* (New York: Routledge, 2006), pp. 127–149 and Steve Macek, *Urban Nightmares: The Media, the Right, and the Moral Panic over the City* (Minneapolis: University of Minnesota Press, 2006).
44. "Zoot Suit Riots," *Encyclopedia Britannica*. www.britannica.com/event/Zoot-Suit-Riots, accessed August 25, 2015.
45. Jonathan Coleman, *Long Way to Go: Black and White in America* (New York: Atlantic Monthly Press, 1997), p. 23.
46. www.ucr.fbi.gov/collections/hate-crime/2014, accessed August 21, 2015 and Southern Policy Law Center (www.splcenter.org) for listings of actual hate crimes.
47. 24/7 Wall Street website, accessed August 18, 2015, www.247wallst.com/special-report/2014/11/18/10-worst-cities-for-lbgt-rights, and Eric Lightblau, "Level of Hate Crimes against Muslims Highest since 9/11," *New York Times*, September 18, 2016.
48. "Hate Crimes and Violence against People Experiencing Homelessness," Coalition for the Homeless website, posted 2012, accessed August 30, 2015, www.nationalhomeless.org/factsheets/hatecrimes.html, and Gina Bellafante, "The Dark Ages of Giuliani," *New York Times*, August 30, 2015.
49. Christopher Ingraham, "Anti-Muslim Hate Crimes Are Still Five Times More Common Today than before 9/11," *Washington Post*, February 11, 2015.
50. Daniel Denvir, "Dearborn: Where Americans Come to Hate Muslims,"

posted 2012, accessed August 12, 2015, www.citylab.com/politics/2012/09/dearborn-where-americans-come-to-hate-muslims/3.

51. "Anti-Muslim Sentiment Grows in US after Paris Attacks," RT America website, accessed January 20, 2016, http://www.rt.com/usa/323148-muslim-attacks-violence-america; Jessica Mendoza, "Amid Anti-Muslim Backlash in US, a Call for Compassion," *Christian Science Monitor*, November 18, 2015; Jenna Johnson, "Trump's Rhetoric on Muslims Plays Well with Fans, but Terrifies Others," *Washington Post*, February 29, 2016; and Cathy Lee Grossman, "Poll: Americans Fear Terrorism, Mass Shootings—and Often Muslims as Well," *USA Today*, December 10, 2015.
52. Iris Marion Young, *Responsibility for Justice* (Oxford: Oxford University Press, 2011).
53. David L. Horowitz, *Ethnic Groups in Conflict* (Berkeley: University of California Press, 1985).
54. John Marzulli, "White Firefighters Rally at Brooklyn Federal Court over FDNY Hiring," *New York Daily News*, October 2, 2012.
55. On Milwaukee, see "Who Gets Construction Jobs and Where?" Employment and Training Institute, University of Wisconsin–Milwaukee, accessed August 21, 2015, www.eti.uwm.edu. On national-level occupation data, see "Labor Force Statistics from the Current Population Survey," accessed August 21, 2015, www.bls.gov/cps/cpsaat11.htm. See also as regards the construction trades, Averil Morrison, Janenne Gonzalez, and Delisa Jones, "Old-Boy Network Is Keeping Our Unions Overwhelmingly Male and White," *Crain's New York Business* 32, no. 19 (2016), p. 13.
56. John R. Logan and Brian J. Stults, "The Persistence of Segregation in the Metropolis: New Findings from the 2010 Census," US2010 Project (New York: Russell Sage Foundation, 2011), p. 2. See also Douglas S. Massey and Nancy A. Denton, *American Apartheid: Segregation and the Making of the Underclass* (Cambridge, MA: Harvard University Press, 1993), quote on p. 9. See also "Latinos Suffer Housing Discrimination in 3 Southern Cities," Fox News Latino, accessed January 20, 2016, http://foxnews.com/latino/news/2013/07/22/latinos-suffer-housing-discrimination-in-3-southern-cities.
57. Kenneth T. Jackson, *Crabgrass Frontier: The Suburbanization of the United States* (New York: Oxford University Press, 1985), pp. 195–218. On St. Louis, see Colin Gordon, *Mapping Decline: St. Louis and the Fate of the American City* (Philadelphia: University of Pennsylvania Press, 2008).
58. Logan and Stults, "The Persistence of Segregation in the Metropolis." On the intersection of segregation and poverty, see Matthew Desmond, *Evicted: Poverty and Profit in the American City* (New York: Crown Publishers, 2016).
59. A listing of fair housing complaints and law suits can be found at Fair Housing Coach website, accessed August 22, 2015, www.fairhousingcoach.com/category/online-alerts/cases-and-settlements. The first federal fair housing legislation was Title VIII of the Civil Rights Act of 1968.

60. Bryce Covert, "How a Poor Neighborhood Becomes a Trap," ThinkProgress website, posted 2015, accessed January 20, 2016, http://thinkprogress.org/economy/2015/08/14/concentrated-poverty/; Peter Dreier, John Mollenkopf, and Todd Swanstrom, *Place Matters: Metropolitics for the Twenty-First Century* (Lawrence: University Press of Kansas, 2001), pp. 56–91; and Robert J. Sampson, *Great American City: Chicago and the Enduring Neighborhood Effect* (Chicago: University of Chicago Press, 2012).
61. Richard Sennett, *Respect in a World of Inequality* (New York: W. W. Norton, 2003).
62. Richard Sennett, *The Fall of Public Man* (New York: W. W. Norton, 1992 [orig. 1974]), p. 255.
63. Howard Gillette Jr., "The City in American Culture," in *American Urbanism*, ed. Howard Gillette Jr. and Zane L. Miller (New York: Greenwood Press, 1987), pp. 27–47; Raymond Williams, "Metropolitan Perceptions and the Emergence of Modernism," *The Politics of Modernism* (London: Verso, 1989), pp. 37–48; and Rosalyn Deutsche, "Men in Space," *Strategies* 3 (1990):130–137. The quotation is from Warren Magnusson, *Politics of Urbanism: Seeing Like a City* (London: Rout ledge, 2011), p. 32. See also Walzer, *On Toleration*, p. 9.
64. Thomas Bender, "City Lite," *Los Angeles Times*, December 22, 1996. On the "theme park" concern, see Michael Sorkin, ed., *Variations on a Theme Park: The New American City and the End of Public Space* (New York: Noonday Press, 1992). On the problematic diversity of public spaces, see Alexander J. Reichl, "The High Line and the Ideal of a Democratic Public Space," *Urban Geography* 37, no. 6 (2016):904–925.
65. Wilkerson, *The Warmth of Other Suns*, p. 45 and Kevin Boyle, *Arc of Justice: A Saga of Race, Civil Rights, and Murder in the Jazz Age* (New York: Henry Holt, 2004).
66. Paul Goldberger, "Atlanta Is Burning," *New York Times Magazine*, June 23, 1996, pp. 52–55.
67. Alison Maney, "These are the Transgender Bathroom Wars, in a Nut Shell," posted 2016, accessed April 23, 2016, www.huffingtonpost.com/kicker/these-are-the-transgender_b_9752266.html.
68. Herbert Gans, *The Urban Villagers: Group and Class in the Life of Italian Americans* (New York: Free Press, 1962).
69. LaFraniere, Cohen, and Oppel, "How Often Do Mass Shootings Occur?" Mass shootings are those that have four or more victims.
70. Richard Faussert and Alan Blinder, "Era Ends as South Carolina Lowers Confederate Flag," *New York Times*, July 10, 2015.
71. Public figures frequently find themselves being publicly criticized and even fired from their positions or forced to resign for intemperate speech. For a recent and typical example, see Richard Sandomir, "ESPN Finally Grows Tired of Schilling's Language," *New York Times*, April 22, 2016.
72. On the connection between wealth and tolerance, see Richard Florida,

The Rise of the Creative Class (New York: Basic Books, 2002). Florida argues that creative people are essential to the economic growth of cities and that they gravitate to places that are "diverse, tolerant and open to new ideas" (p. 249). See also Richard Florida and Gary Gates, "Technology and Tolerance," Center for Urban and Metropolitan Policy (Washington, DC: The Brookings Institution, 2001).

73. John Elgin and Richard Fausset, "Defiant Showing of Unity in Charleston Church That Lost 9 to Racist Violence," *New York Times*, June 21, 2015.

CHAPTER SIX

1. On San Francisco see Daniel Goldstein, "San Francisco Real Estate Looks like It Did before Dot-Com Crash in 2000," *MarketWatch*, posted 2016, accessed February 18, 2016, at www.marketwatch.com/story/san-francisco-real-estate-looking-like-it-did-before-dotcvom-crash-in-2000-2015-11-20. As regards New York City, see Martin Filler, "New York: Conspicuous Construction," *New York Review of Books*, April 2, 2015; Robert Frank, "If It's March, They Must Be in Miami," *New York Times*, March 3, 2016; and Daniel Geiger, "Comeuppance," *Crain's New York Business* 32, no. 11 (2016):12–13. Matthew Desmond documents the other extreme of the housing market in his *Evicted: Poverty and Profit in the American City* (New York: Crown Publishers, 2016).
2. See Ichiro Kawachi, Bruce P. Kennedy, and Kimberly Lochner, "Long Live Community: Social Capital and Public Health," *American Prospect*, November/December 1997, pp. 56–59 and Richard Wilkinson, *Unhealthy Societies: The Afflictions of Inequality* (London: Routledge, 1996).
3. Elizabeth Svoboda, "America's 50 Greenest Cities," *Popular Science*, accessed June 7, 2015, www.popsci.com/article/2008–02/americas-50-greenest-cities, and Susan Brady, "The Most Toxic Cities in America," *HealthNews*, posted 2015, accessed March 6, 2016, www.healthnews.com/en/news/The -Most-Toxic-Cities-in-America/3SOCx8cdrBlws/Eb6roxWwo.
4. "Flint Water Crisis Fast Facts," CNN Library, accessed March 12, 2016, www.cnn.com/2016/03/04/us/flint-water-crisis-fast-facts.
5. In a recent article on "The 6 Most Corrupt Cities in American, *AvvoStories* listed four shrinking cities (Detroit, Newark, Philadelphia, and New Orleans) along with Las Vegas and Chicago. And even though Chicago is often described as prosperous, it has lost significant population since the 1950s. http://stories.avvo.com/money/the-6-most-corrupt-cities-in-America.html, accessed March 15, 2016.
6. Bryce Covert, "How a Poor Neighborhood Becomes a Trap," *ThinkProgress*, posted August 14, 2015, accessed January 20, 2016, www.thinkprogress.org/economy/2015/08/14/3691749/concentrated-poverty/.
7. "Fighting Police Abuse: A Community Action Manual," American Civil

Liberties Union website, accessed March 15, 2016, www.aclu.org/fighting-police-abuse-community-action-manual.

8. Richard Florida, "Tolerance and Intolerance in the City," CityLab website, posted 2015, accessed March 6, 2016, www.citylabs.com/housing/2015/05/tolerance-and-intolerance-in-the-city/304.
9. Wilkinson, *Unhealthy Societies*, p. 5.
10. John Friedmann, *Planning in the Public Domain* (Princeton, NJ: Princeton University Press, 1987), pp. 343–388 and James Holston and Arjun Appadurai, "Cities and Citizenship," in *Cities and Citizenship*, ed. James Holston (Durham, NC: Duke University Press, 1999), pp. 1–18.
11. On rights and responsibilities as they pertain to urban citizenship, see Robert A. Beauregard and Anna Bounds, "Urban Citizenship," in *Democracy, Citizenship and the Global City*, ed. Engin F. Isin (London: Routledge, 2000), pp. 243–256. See also Iris Marion Young, *Responsibility for Justice* (Oxford: Oxford University Press, 2001).
12. Regarding "attention to the urgent problems of the moment" (p. 228), see George Orwell, "Inside the Whale," in George Orwell, *A Collection of Essays by George Orwell* (New York: Harcourt Brace Jovanovich, 1946 [orig. 1940]), pp. 210–252.
13. Jeffrey C. Goldfarb, *The Cynical Society: The Culture of Politics and the Politics of Culture in American Life* (Chicago: University of Chicago Press, 1991).
14. Young, *Responsibility for Justice*, p. 52.
15. Iris Marion Young, *Justice and the Politics of Difference* (Princeton, NJ: Princeton University Press, 1990).
16. The quotations in the paragraph are from Catherine Fennell, *Last Project Standing: Civics and Sympathy in Post-Welfare Chicago* (Minneapolis: University of Minnesota Press, 2015), pages 8 and 251 respectively. On the acceptance of difference, see Young, *Justice and the Politics of Difference*, pp. 156–191. As for respect, see Avishai Margalit, *The Decent Society* (Cambridge, MA: Harvard University Press, 1996). Lastly, regarding allowances for others and interceding when people transgress, many gray areas exist and what to do is not obvious. On this point, see Nancy L. Rosenblum, *Good Neighbors: The Democracy of Everyday Life in America* (Princeton, NJ: Princeton University Press, 2016).
17. Young, *Responsibility for Justice*, pp. 95–122.
18. On this point, see Carlo Rotella, *October Cities: The Redevelopment of Urban Literature* (Berkeley: University of California Press, 1998), pp. 1–16. More generally and more theoretically regarding how we experience and imagine the city, see Ash Amin and Nigel Thrift, *Cities: Re-imagining the Urban* (Cambridge: Polity, 2002).
19. Gary Bridge, *Reason in the City of Difference* (Oxon, UK: Routledge, 2005), pp. 15–38. Much of the writing on this topic concerns gendered bodies, sexuality, danger, and desire. See Fran Tonkiss, *Space, the City and Social Theory* (Cambridge: Polity, 2005), pp. 94–112.

20. Numerous authors have explored these experiences by "walking the city." For two recent examples, see Teju Cole, *Open City* (New York: Random House, 2011) in the genre of fiction and Vivian Gornick, *The Odd Woman and the City* (New York: Farrar, Straus, and Giroux, 2015) in nonfiction.
21. The reference here is to the famous essay by Georg Simmel titled "The Metropolis and Mental Life." See *The Culture of Cities*, ed. Richard Sennett (New York: Appleton-Century-Croft, 1969), pp. 47–60.
22. Robert A. Beauregard, *Voices of Decline: The Postwar Fate of U.S. Cities* (New York: Routledge, 2003) and Steve Macek, *Urban Nightmares: The Media, the Right, and the Moral Panic over Cities* (Minneapolis: University of Minnesota Press, 2006). As regards social media, see "Social Media and the City," IBM Smarter Cities website, accessed April 26, 2016, http://www.ibm.com/smarterplanet/us/en/smarter_cities/overview/.
23. On representation, see James Donald, *Imagining the Modern City* (Minneapolis: University of Minnesota Press, 1999). On the phantasmagoria of city life, see Steve Pile, *Real Cities* (London: Sage, 2005). The classic book on spatial imaginaries is Benedict Anderson, *Imagined Communities* (London: Verso, 1983). For a specific example involving New York City, see May Joseph, *Fluid New York: Cosmopolitan Urbanism and the Green Imagination* (Durham, NC: Duke University Press, 2013).
24. Tod Newcombe, "The City that Incorporated Social Media into Everything," *Governing*, posted 2015, accessed April 26, 2016, www.governing.com/columns/tech-talk/go-integrating-social-media-roanoake.html.
25. Mike Davis, *Ecology of Fear: Los Angeles and the Imagination of Disaster* (New York: Metropolitan Books, 1998). More generally on the city in fiction, see Richard Lehan, *The City in Literature: An Intellectual and Cultural History* (Berkeley: University of California Press, 1998). And, for a specific city: Barbara Eckstein, *Sustaining New Orleans: Literature, Local Memory, and the Fate of the City* (New York: Routledge, 2006).
26. Atticus Lish, *Preparation for the Next Life* (New York: Tyrant Books, 2015), p. 83.
27. Bridge, *Reason in the City of Difference*; Warren Magnusson, *Politics of Urbanism: Seeing Like a City* (London: Routledge, 2011).
28. Andreas Huyssen, "World Cultures, World Cities," in *Other Cities, Other Worlds*, ed. Andreas Huyssen (Durham, NC: Duke University Press, 2008), pp. 1–23.
29. Matthew Gandy, *The Fabric of Space: Water, Modernity, and the Urban Imagination* (Cambridge, MA: MIT Press, 2014).
30. Stephen Graham and Simon Marvin, *Splintered Urbanism: Networked Infrastructures, Technological Mobilities, and the Urban Condition* (London: Routledge, 2001).
31. Hydraulic fracking involves using water and chemicals to extract oil from underground rock formations. The waste water from the process is then disposed of beneath the aquifer. In Oklahoma and other states, this has

dramatically increased the number and severity of earthquakes. See William Yardley, "Quakes Finally Jolting Oklahoma Officials to Act—Increase in Seismic Activity Is Seen as a Result of Fracking," *Los Angeles Times*, March 2, 2015 and Pamela Worth, "Got Science? Will Oklahoma Finally Get Serious about Fracking-Related Earthquakes in 2016," *Huffington Post*, posted 2016, accessed April 28, 2016, www. huffingtonpost.com/pamela-worth/got-science-will-oklahoma b 9195262.html.

32. Magnusson, *Politics of Urbanism*, p. 117.

Index